THE LIBRARY
P9-AFZ-582

OIL AND THE INTERNATIONAL ECONOMY

OIL AND THE INTERNATIONAL ECONOMY

BY

Geoffrey Heal

and

Graciela Chichilnisky

CLARENDON PRESS · OXFORD
1991

Oxford University Press, Walton Street, Oxford OX2 6DP
Oxford New York Toronto
Delhi Bombay Calcutta Madras Karachi
Petaling Jaya Singapore Hong Kong Tokyo
Nairobi Dar es Salaam Cape Town
Melbourne Auckland
and associated companies in
Berlin Ibadan

Oxford is a trade mark of Oxford University Press

Published in the United States
by Oxford University Press, New York

British Library Cataloguing in Publication Data
Heal, Geoffrey
Oil and the international economy.
1. Petroleum industries. International economic aspects
I. Title II. Chichilnisky, Graciela
338.27282
ISBN 0–19–828517–5

Library of Congress Cataloging in Publication Data
Heal. G. M.
Oil and the international economy/by Geoffrey Heal and Graciela Chichilnisky.
p. cm.
Includes bibliographical references and index.
1. Petroleum industry and trade. 2. International economic relations. I. Chichilnisky, Graciela. II. Title.
HD9560.5.H43 1991 382'.42282—dc20 90–40416
ISBN 0–19–828517–5

Typeset by Cambrian Typesetters, Frimley, Surrey
Printed in Great Britain by Biddles Ltd
Guildford and King's Lynn

Contents

List of Text Figures

List of Tables

Preface

This book is about the economic institutions through which oil is produced, traded, and consumed, and about their interactions with national and international economic systems. Oil plays a unique role in meeting certain aspects of the energy demands of the industrial countries, and is likely to continue to do so well into the next century. As long as this is the case, the oil market will occupy a central role in the international economy, and in the economies of many producing and consuming countries. To understand this market well, one has therefore to see these interactions clearly, and to grasp how the oil market relates to the world's economic and financial systems, and to the economies of industrial and developing countries. Only then is it possible to understand the economic forces that are driving the international oil market.

A particularly important feature of the oil market is the complexity of the network of interdependence that it generates. Industrial and developing countries, planned and market economies, producing and consuming countries, industrial and financial systems: all are affected by developments in the oil market. This market exemplifies the growing interdependence of the world economy, and provides an instructive case-study of its origins and its implications.

It is not only the international linkages of the oil market that are important: it is also a matter of great significance that one of the major actors in this market, OPEC,[1] is an association of non-industrial countries. The emergence of OPEC to a position of world-wide economic influence was an omen of emerging changes in international economic relations. The rise in oil prices of the 1970s effected a major international redistribution of wealth. This has been substantially from rich to poor, and on a scale which could never have been achieved through official development assistance.[2] This example of self-help on the part of developing countries has inspired several attempts at imitation, and has also led to changes in the political climate of North–South relations, changes which are being reinforced by other more gradual, and hence less noteworthy, developments in the international economy.

Perhaps as much as anything because of its role as a portent of possible changes in the balances of economic power, the oil market has become a source of controversy and of misconception. OPEC has been blamed for many of the economic ills of both industrial and developing countries. There have been pronouncements of an almost religious nature by economists and politicians about the harmful effects of high oil prices, and about the proper role of market forces in the oil market. For an outsider, it has become difficult to separate conventional wisdom and wishful thinking from reality on many of these issues.

To anticipate just a few conclusions from the following chapters:

—higher oil prices have not had a major impact on growth or inflation rates in the industrial economies (Chapter 6).

—even without OPEC, the price of oil would have risen in the 1970s (Chapter 1).

—high energy prices did not have a substantially harmful effect on oil-importing developing countries (Chapter 8).

—oil-producing countries did not gain as much as is widely believed from higher oil prices (Chapter 7); they often lost more than they gained.

In the following chapters we present an analytical framework for studying the basic economic forces at work in the world oil market, and then apply this framework to a number of issues, including those just alluded to above. The conclusions that emerge are sometimes surprising.

This summary makes it clear that this is not a book about the oil industry, or about the oil market alone. It addresses basic structural issues in the oil market, and then sets this market in the broader context of the international economy, focusing on its macroeconomic implications and on its impact on patterns of international trade. There are many excellent studies of the oil industry and of the oil market, and we seek to complement rather than compete with these.[3]

An extensive technical literature on resource markets has developed since the early 1970s, building on the work of Hotelling.[4] This tradition focuses on the exhaustibility of resources, and on their allocation over time. It asks, and answers, such questions as: 'How fast should we deplete resources stocks?' 'Will markets deplete resource stocks at the correct rate?' 'Is there a systematic tendency for markets to deplete resource stocks too fast?' We review a limited part of this body of theory in a non-technical way in Chapters 1 and 2, to provide a background for understanding the role of OPEC and of other forces in determining the movement of oil prices over time.[5] This theory, and other considerations, provide a basis for our analysis of the role of OPEC in Chapter 4 and our analysis of the evolution of the world oil market in Chapter 5.

There is a different dimension to energy use, an atemporal one which focuses on the distributional and macroeconomic consequences of energy use, rather than on dynamic aspects. This approach has a long pedigree, as this is the framework within which the classical economists viewed natural resource use. They studied the implications of resource endowments for the distribution of income and wealth. In the last two decades this tradition has received less emphasis than the dynamic tradition based on analysis of exhaustion. We place considerable emphasis on it here. Many aspects of the impact of the world oil market on the rest of the international economic system cannot be understood without a proper grasp of these distributional and macroeconomic considerations.

In studying the macroeconomic and distributional consequences of changes in the world oil market, we use a general equilibrium approach. There are two approaches to the study of resource allocation problems such as these: partial and general equilibrium approaches. In the partial equilibrium approach we study what happens in one market, assuming that the changes in this market can be isolated and that they have no effects on other parts of the economy which could feed back to the market we are studying.

Allowing for the full range of interactions between the market under study and the rest of the economy is the hallmark of general equilibrium analysis. Its relevance in the present context is that a change in the price of a good as important as oil is bound to have an impact on a wide range of other prices, which in turn will feed back to the oil market. For example, a change in the nominal price of crude oil will affect the price of goods and services produced by oil importers, which are frequently the imports of oil exporters. Hence the costs of the oil producers are affected by the price which they set for oil. Feedbacks of this type necessitate a general equilibrium approach, which is a distinctive feature of the second half of this book.[6]

This approach enables us to provide simple but rigorous explanations of several apparently paradoxical empirical findings. For example, empirical studies indicate that capital and oil are sometimes complements and sometimes substitutes. General equilibrium feedbacks can explain both relationships within the same model, predicting substitutability at certain price ranges and complementarity at others. This has important policy implications and is analysed in Chapter 3. Our model can also explain why some oil exporters have prospered from their oil-exporting activities, while others have found oil to be a double-edged sword, with oil exports leading to foreign debt, payments deficits and the destruction of the agricultural sector. This is the subject-matter of Chapters 7 and 8. All of these issues have been studied before, but without an integrated treatment being offered. Our treatment integrates the study of oil exports within a more general framework for the study of commodity exports, although oil remains unique among commodities in that it faces a very inelastic demand and plays an important role in the industrial economies. In sum, we can integrate the oil question into general issues of North–South trade, while reconciling apparently conflicting empirical findings.

The material is presented in a non-technical way, in order to make it accessible to anyone with an interest in the field, whatever may be their background. Inevitably this means that on occasions the reader seeking a completely detailed development of the argument will be referred elsewhere. This should not be a problem: technical arguments are well covered in the literature, but what has been lacking is an application of the existing technical economic analysis to an overview of the issues that are of general concern in the context of the oil market.

This book has developed via a series of courses taught to graduate students at Essex University, at the Woodrow Wilson School, Princeton University, and at the Graduate School of Business, Columbia University. Their almost insatiable appetites for discussion of the world oil market have convinced us that our own interests are widely shared, and that there is a real role for a short, analytical, less technical review of the issues.

The research underlying the book was funded by the United Nations Institute of Training and Research (UNITAR), the Rockefeller Foundation, and the Faculty Research Fund of the Columbia Business School, and we are grateful to them, and in particular to Dr Phillipe de Seynes and Mr Michael Bloome, for their enthusiastic encouragement. Discussions with Partha Dasgupta have greatly improved our understanding of the very complex material that we have had to digest in writing the book: his own researches have often been instrumental in developing important insights, and are frequently referred to below. We would also like to acknowledge the very valuable contribution of research assistance from Rachel Pohl and Anirudha Basu.

NOTES

1. The Organization of the Petroleum Exporting Countries.
2. Normally referred to as foreign aid.
3. A classic is M. A. Adelman, *The World Petroleum Market* (1972). A book with a more limited focus is J. M. Griffin and D. J. Teece, *OPEC Behaviour and World Oil Prices* (London, 1982).
4. Harold Hotelling, 'The Economics of Exhaustible Resources', *Journal of Political Economy* (1931), vol. 39, pp. 137–75.

5. A comprehensive and detailed review is given in P. S. Dasgupta and G. M. Heal, *Economic Theory and Exhaustible Resources* (Cambridge, 1979). Less technical reviews can be found in J. M. Hartwick and N. D. Olewiler, *The Economics of Natural Resource Use* (1986), or in J. M. Griffin and H. B. Steel, *Energy Economics and Policy* (1986).
6. Relevant references are Chichilnisky, 'A General Equilibrium Theory of North–South Trade', Ch. 1, vol. II in Heller, Starr and Starret (eds.), *Essays in Honour of Kenneth J. Arrow* (Cambridge, 1988); Chichilnisky, 'International Trade in Resources: a General Equilibrium Analysis', in R. McKelvey (ed.), *Environmental and Natural Resource Mathematics*, *Proceedings of Symposia on Applied Mathematics*, (American Mathematical Society Providence, Rhode Island, 1986), pp. 75–125; and Chichilnisky, 'Prix du Petrole, Prix Industriels et Production: une Analyse Macroeconomique d'Equilibre General', in G. Gaudet and P. Laserre (eds.), *Resources Naturelles et Theorie Economique* (Quebec, 1986), pp. 26–56.

1
Competition and Oil Prices

1.1 SCARCITY AND OIL PRICES

Oil is scarce. And if the rate of consumption exceeds the rate of discovery, as is currently the case, it becomes scarcer. What is the consequence of a commodity becoming scarcer with the passage of time? If demand stays roughly constant, then its price can be expected to rise. As the consumption rate exceeds the discovery rate, the supply of oil falls relative to demand. Oil becomes scarcer, and the oil market is equilibrated by a rise in price.

This simple point is crucial to an understanding of the long-run behaviour of the world oil market. Oil is becoming scarcer, so that with constant demand conditions its price can be expected to rise. This has nothing to do with OPEC, with the major oil companies, or with monopoly or extortion. It is the outcome of elemental economic forces. Of course, monopoly might contribute to the price rise, but there is no need to invoke growing monopolization to explain the long-run rising trend in oil prices. By analogy, as population in an area increases, land prices increase, because demand for land rises relative to a fixed supply. This is not evidence of monopolization of the land market, but of the changing balance between supply and demand.

In fact, in the oil market, in the short run, this balance could move either way: a recession and the associated drop in demand could interrupt the long-run trend, temporarily decreasing scarcity. New discoveries of oil deposits can also reverse this trend, as in the 1950s and 1960s, a period during which there was a glut of oil and prices fell substantially (see Chapter 5). New technologies can also relieve the growing pressure of scarcity, and indeed an effective 'backstop' technology (see Chapter 2) could halt it altogether. But no such technology is available yet, in spite of billions of dollars having been spent on nuclear energy, synfuels, etc. It is clear, then, that underlying any perturbations, there is currently a long-run trend towards scarcity. Although the potential of growing scarcity is economically the most distinctive feature of the oil market, it is not always perceived as such. Public perception of scarcity is cyclical. It peaked in the 1970s and is now undergoing a revival.

So perhaps the most important point that economic theory can make about the oil market is that in the long run prices can be expected to rise, until technological developments provide an effective long-term substitute for oil or demand conditions change radically.

This observation can be formalized in the following terms. Call the cost of extracting one extra barrel of oil the 'marginal extraction cost' (MEC) of oil. The market price of oil minus its MEC will be called its 'net price'. This is the extra profit to a producer from producing and then selling an extra barrel. *Then in a competitive market, with all traders fully informed about present and future prices, the net price of oil (market price minus MEC) will rise at a rate equal to the rate of interest.* In other words, the market price can be expected to draw away from the marginal extraction cost at a rate equal to the interest rate. This result is known as 'Hotelling's Rule', after its first discoverer.[1]

In concrete terms, suppose the current market price to be \$30 per barrel, with the MEC \$16 per barrel and the interest rate 10 per cent. Then net price is \$14 per barrel, and this will rise 10 per cent per year, giving a net price of \$15.4 per barrel one year hence. So if the MEC stays constant at \$16, the market price one year ahead should be \$31.4 per barrel, an increase of \$1.4 on the present \$30 price.

This conclusion, that net price raises at the interest rate in a competitive market, is reached by analysing the behaviour of profit-maximizing producers of oil. This analysis, as we shall see, rests on a number of premisses which are palpably not good descriptions of reality, but it is nevertheless an important benchmark. It formalizes and quantifies the idea that as oil becomes scarce because reserves are depleted, market forces will drive up the price. It enables us to work out, for a particular conjunction of circumstances, just how much the price will rise.

While it is clear that growing scarcity will send the price up, the role of the interest rate is not so immediate. Why should the net price rise at a rate equal to the interest rate, rather than at some other rate?

To see this, note first that if the price net of costs rises at the interest rate, then in fact the *present value* of this net price is the same at all dates. Another way of saying this is the following: suppose that you can sell your oil for a profit of 1 dollar today. If the net price of oil increases at the interest rate, then you can produce and sell your oil at any future date and obtain the same sum of money as you would have obtained from selling today and investing your dollar.

More technically, the present discounted value of the profits from producing and selling an extra barrel of oil are the same whether the sale is now, next year, or later, provided that the net price of oil rises at the interest rate.

This condition, that the profits from an extra barrel are the same (in present value terms) in all years, is in fact necessary for competitive markets to clear. To see this, suppose to the contrary that the present value profits from producing and selling an extra barrel were in fact larger next year than in any other year. Then any profit-maximizing producer would prefer to sell next year rather than this, and would therefore delay production and sales until next year. There would be a current shortage as sellers postpone production and sales, and a glut next year when all the postponed output reaches the market. The oil market would not be in equilibrium: demand and supply would not be matched up at each date.[2]

This theorizing tells us not only that oil prices will rise over time, but also that they will do so at a rate depending on the rate of interest. If these arguments can be shown to have practical relevance, then this is an important insight. Interest rates world-wide are influenced by the macroeconomic policies of the industrial countries, so that these policies may have an impact on oil price movements.

In fact the relationship between interest rates and oil prices is rather more complex than indicated so far. It has been established that under certain conditions the net price will rise at the rate of interest. But what happens if, for example, there is an unanticipated, once-and-for-all, rise in the rate of interest? Obviously, after the rise, the rate of increase of the net price will equal the new higher value. But there is in addition to this a once-and-for-all drop in the current price level. So the consquences of an unanticipated rise in the rate of interest are a sharp drop in the price level, followed by growth at a higher rate than before. This is shown in Fig. 1.1.

Why should there be this once-and-for-all drop in the net price of oil when interest rates rise? The easiest way to see this is to note that oil reserves are an asset, just like government bonds or corporate stocks. Now we know what happens to the prices of these assets when interest rates rise. It is a matter of general observation that they fall. They fall because any returns that they may offer—dividends, profits, or capital gains—are now less attractive relative to the returns on interest-bearing assets. So the value of oil as an asset will also fall as the interest rate rises, leading to a drop in its price. There is a new lower level of oil prices, but of course a higher rate of increase from then on. The interest rate policies of the oil-consuming

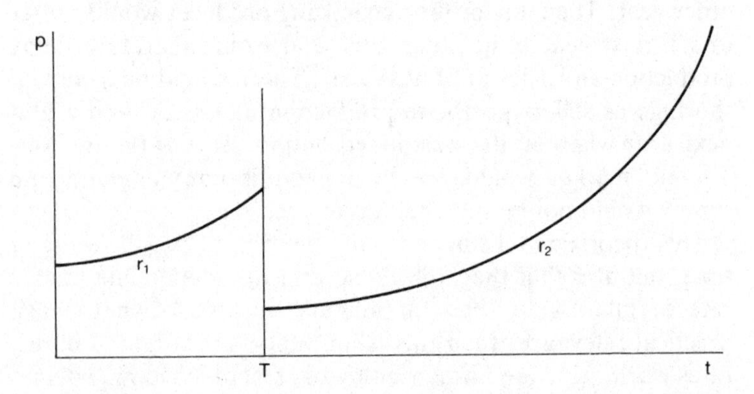

Fig. 1.1: The time path of the net price of oil in a competitive market if there is an unanticipated rise in the interest rate at date T from r_1 to r_2.

countries can therefore have a very complex effect on oil prices: this is an interesting dimension of the interdependence of consumer and producer countries via the world oil market. The level of interest rates determines the rate of change of oil prices, and a change in interest rates will alter both the level and the rate of change of oil prices.

As an illustration, consider the effect of the increase in interest rates in the industrial countries in the late 1970s and early 1980s. The theory suggests that this could have contributed to the drop in oil prices of the early 1980s. It also suggests that once interest rates stabilized at the new higher level, and the short-run effect worked through giving a lower level of oil prices, the price of oil should start rising again at a higher rate than before. Later chapters will consider in more detail forecasts for the oil market, and we shall review the evidence on the relationship between oil prices and interest rates later in this chapter.

In summary, one important point that we quickly appreciate from this economic analysis is that in the long-run the price of oil will rise, given constant demand conditions (i.e. no major changes in the structure of oil use) and constant supply conditions. Furthermore, price increases can be related to interest rates in oil-consuming countries.

Another important implication of this simple theorizing is that the price of oil is not tied to its marginal extraction cost (MEC), even under competitive conditions. It is often asserted that in the long run the competitive price of oil must equal its MEC, and that an increase in competition in the oil market would drive it down to this level. Conversely, it is taken as evidence of the exercise of monopoly power that the price of oil exceeds its MEC. These assertions are simply wrong for oil. Even in a competitive market, its price will draw steadily away from its cost. *A price–cost gap is no evidence of monopoly power, nor is a growing price–cost gap evidence of growing monopoly power*. Both are facts with which we would have to live even in a perfectly competitive world.

What would determine the net price, i.e. the price–MEC gap, in a competitive market? This would depend upon the remaining reserves of oil, and upon present and future demand conditions. There are ways of quantifying this dependence theoretically, but it is hard to obtain reliable data to evaluate these theoretical relations. So whether the present net price is 'right', or is 'too big', is an issue that is not easily resolved.[3] Note that the price–cost gap will not necessarily rise forever: the existence of substitutes for oil which become competitive at high prices may place a limit on sustainable market price levels for oil, and thus on sustainable price–cost gaps.[4] However, even with abundant and competitive substitutes the price–cost gap may remain positive. Incidentally, the net price, i.e. price minus MEC, is often referred to as 'royalty' or 'rent'.

1.2 MONOPOLY AND OIL PRICES

The previous section may have left the reader with the impression that monopoly would not make a qualitative difference to the behaviour of oil prices. Certainly it showed that simple features such as rising prices or prices in excess of costs are not evidence of monopoly. What then is the impact of monopolistic or cartel behaviour in the oil market, and how could it be recognized? In view of the widespread presumption that OPEC exercises, or has exercised, some degree of monopoly power, this question is of obvious interest.[5] It is one to which we shall return at several points in the book, and in particular in Chapter 4. In fact the impact of monopoly on the long-run trend of prices is far from dramatic, and may even be hard to recognize. Its biggest impact may be in a stabilization of prices, a reduction in their variation, rather than a change in their trend behaviour.

An analysis of the consequences of monopoly leads to the conclusion that, relative to a competitive market, a monopolistic oil market will under rather general demand conditions[6] have either:

1. a higher rate of price increase, but a lower initial price, (case a), or
2. a lower rate of price increase, but a greater initial price (case b).

These two possibilities are shown in Fig. 1.2: indeed, a mixture of the two is possible, as in Fig. 1.3. What can definitely *not*

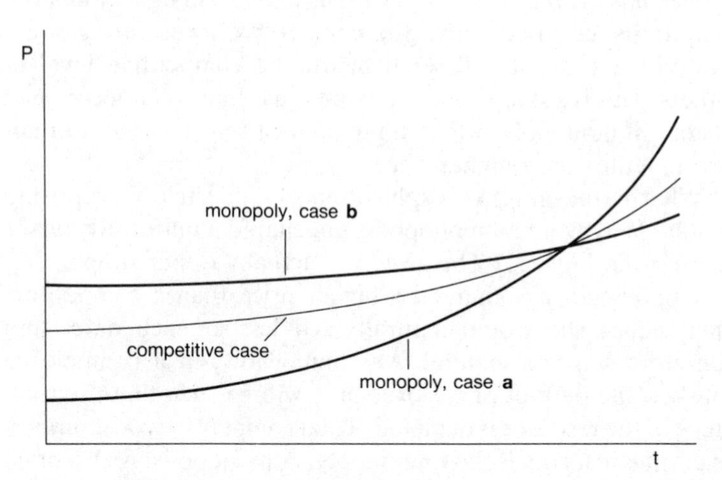

Fig. 1.2: Possible relationships between competitive and monopolistic price paths.

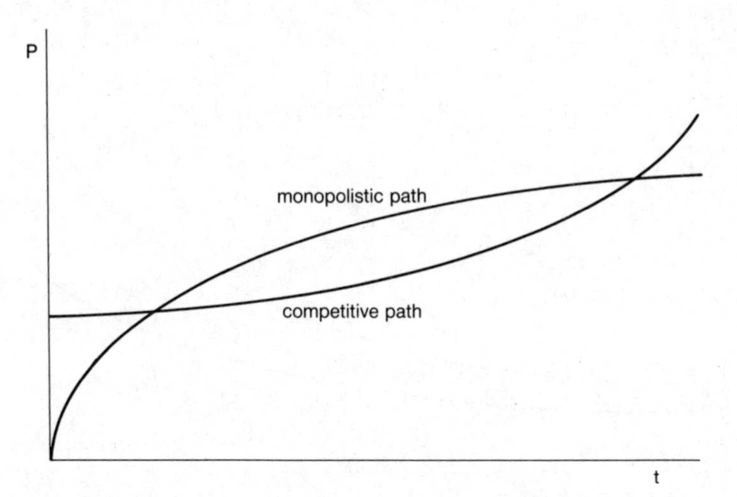

Fig. 1.3: A more complex relationship between monopolistic and competitive price paths.

occur is a monopolistic price path on which price is always higher than the competitive price: Fig. 1.4. is ruled out. *Monopoly in the oil market cannot result in prices uniformly higher than competitive prices at all dates*. Though monopoly may raise the price above the competitive level during some periods of time, it will set it below the competitive level in others. This is a sharp contrast to the usual static microeconomic theory of monopoly, where a monopolist sells less output than a competitor, at a higher price.

What is the intuitive explanation of this slightly surprising result? Why would a monopolist *not* charge a uniformly higher price, as in Fig. 1.4? The answer is actually rather simple. If a monopolist always charged a higher price than a competitor, then he or she would naturally sell less at each date, and therefore sell less in total over time. Now, in a competitive market, the path of prices over time will be such that the total stock of the resource is depleted. Total competitive consumption over time just equals the total supply. A monopolist with a price uniformly higher than the competitive price will therefore find unsold stocks on his or her hands in the long run, because sales are less than competitive sales at all dates and so cumulative sales are less.

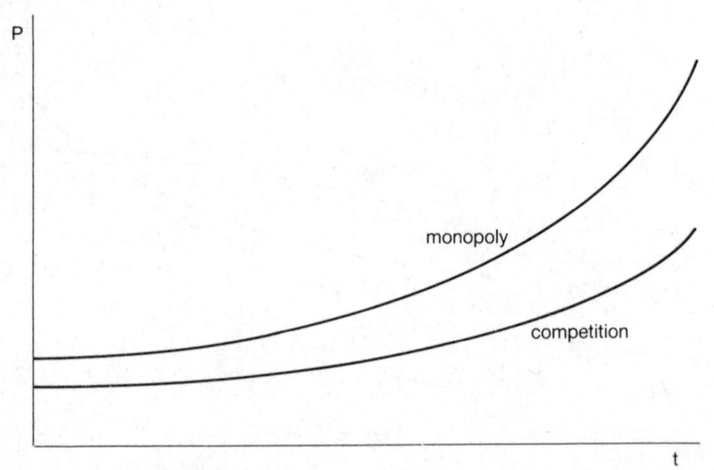

Fig. 1.4: A monopolistic price uniformly above (or below) the competitive case, is not possible.

Consider, to make matters simple, the case where the marginal extraction cost is zero. Then as long as the resource can be sold for a positive price, it is profitable to do so. Therefore, to be left with unsold stocks, stocks which could have been sold at a (positive) price in excess of (zero) marginal extraction cost, is clearly incompatible with maximization of profits. But as a price path uniformly in excess of the competitive price implies unsold stocks in the long run, such a price path is not consistent with profit maximization.

It emerges, then, that a monopolist will not charge a price uniformly higher than that ruling under competition, but will set prices sometimes above and sometimes below the competitive level: this fact is important in interpreting the role of a cartel such as OPEC in the world oil market. This issue will be examined in more detail in Chapter 4.

Is it possible to be more precise about the difference between the competitive and monopolistic cases? In fact it is possible to give a simple analysis of which of the cases shown in Fig. 1.2 will arise. Fig. 1.2 (case a) shows a monopoly price initially below the competitive level, and rising faster: Fig. 1.2 (case b) shows the opposite case. Which of these occurs depends upon demand conditions in the following way. Suppose that at high oil prices, the demand for oil falls away sharply in response to a price increase. Formally, the demand for oil becomes elastic at high prices: a given percentage rise in price causes a bigger proportionate demand response at high rather than low prices.

Such a situation is certainly plausible: as the price of oil rises, there are a number of substitutes which become competitive. Suppose for example that the cheapest synthetic fuel can be produced at $40 per barrel: then as the price of oil passes $40, we can expect demand to drop sharply as some of the market is taken by the new fuel. Alternatively, suppose that as the price of oil passes $38 per barrel, it becomes more profitable to use coal rather than oil as a petrochemical feedstock. Then again there will be a sharp drop in demand as the price of oil passes a critical value, in this case $38 per barrel. In such a world, oil producers will lose their market rapidly as the price of oil rises. This gives them an incentive to avoid high prices, and in particular implies that a rational monopolist would use his or her market power to avoid high prices. Such a monopolist would pick a relatively flat price profile, as in Fig. 1.2 (case b). Cases such as Fig. 1.2 (case b) will thus occur when demand for oil becomes elastic at high prices.

This is an example of what is called 'limit pricing', i.e. setting prices so as to limit entry into a market by possible competitors. It will become clear in the discussion of OPEC in Chapter 4 that the behaviour of OPEC's most powerful members, the Gulf States, fits this pattern well. It is worth remarking that if the price elasticity of demand for oil is greater at higher prices, giving a monopolistic price path such as Fig. 1.2 (case b), then the sudden monopolization of a previously competitive oil market would produce an upward jump in prices to a higher level, but then a lower rate of increase. Fig. 1.5 shows this.

Clearly, monopolistic price paths such as that shown in Fig. 1.2 (case a) (lower initial price, higher rate of increase than the competitive case) will occur under conditions exactly the opposite of those which we have just discussed and which give rise to Fig. 1.2 (case b). They will occur when the demand for oil is more price-elastic at low prices, and becomes less elastic at higher prices.

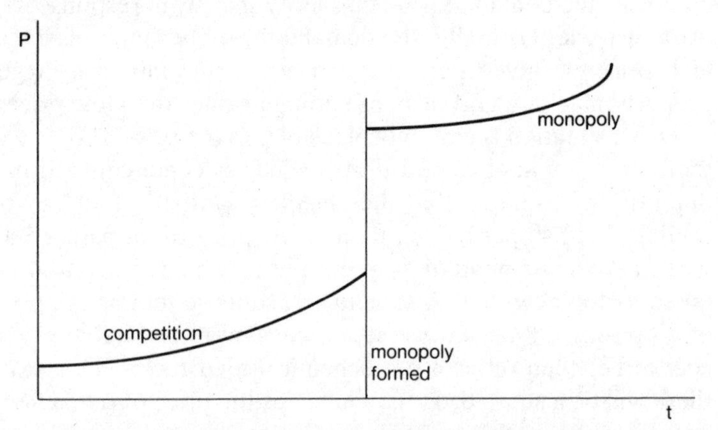

Fig. 1.5: Oil price movements in a market which starts competitive and becomes monopolistic, in the case when price elasticity of demand increases at high prices.

What economic circumstances would give rise to such demand conditions? To appreciate an example of such circumstances, note that a major advantage of oil relative to other fuels is that it has a very high ratio of energy content to weight, and is easy to move. In some fuel uses, such as in transportation, this is crucial. It is difficult to imagine running cars and aircraft on coal. In other uses, such as in electric power generation plants, the energy-to-weight ratio is much less important and it is quite practical to use coal. When the price of oil is low, it competes with coal as an energy source in electric power generation. However, as its price rises, and it becomes more expensive than coal, it will no longer be used in power generation. Its distinctive characteristic, a favourable power-to-weight ratio, is of no value and it is not worth paying extra for this. Transportation users will continue to buy oil as they need this property. Hence the demand for oil as a static, low-grade energy source, a demand which exists at low prices, may be very price-elastic. This would give rise to a monopolistic price path like that shown in Fig. 1.2 (case a), and in this case the monopolization of the oil market would cause a once-and-for-all *drop* in the price of oil, and then a more rapid rate of growth. Fig. 1.6 depicts this case.

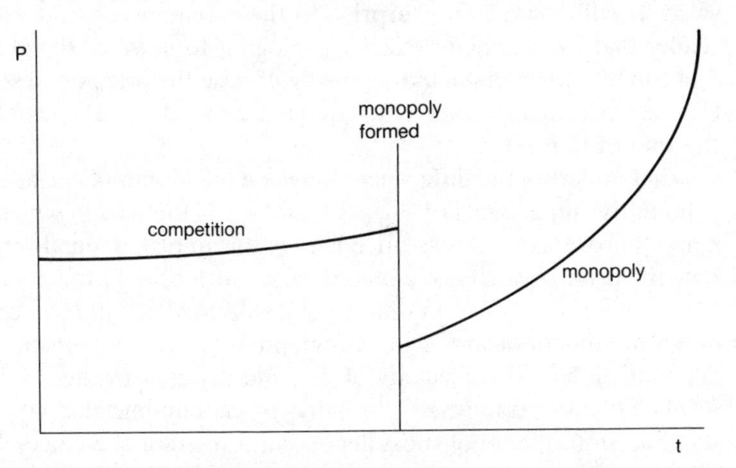

Fig. 1.6: Oil price movements in a market which starts competitive and becomes monopolistic, in the case where the price elasticity of demand increases as the price falls.

Of course, the two classes of demand conditions described above are not mutually exclusive: the demand for oil could be rather elastic at both high and low prices, and less so at intermediate prices. This would lead to monopolistic price paths such as the one shown in Fig. 1.3.

In the first section of this chapter, it was argued that, in a competitive market, the net price of oil would rise at the rate of interest. The net price was defined as market price minus marginal extraction cost, and is of course a measure of the extra profit to be obtained from producing and selling one more barrel of oil. In the case of monopoly, it is necessary to work with a slightly different concept, which however plays an exactly equivalent role. This is the concept of net marginal revenue. Marginal revenue is the increment to a seller's revenue from the sale of one extra unit of a good, taking account of the fact that by putting more on the market, the seller may force the market price down and so receive less revenue on all existing sales. So marginal revenue minus marginal extraction cost—net marginal revenue—is the increment to profits from selling one more unit, for a seller who has enough market power to force the price down by increasing sales. It will come as no surprise to the economically trained reader that by arguments exactly analogous to those of section 1, it can be shown that a monopolist will raise the price of oil so that net marginal revenue (the equivalent of net price) rises at the rate of interest.[7]

What underlies the differences between the monopolistic and competitive price paths of Figs. 1.2 and 1.3 is the way in which marginal revenue varies as price varies: this in turn depends on how the price elasticity of demand varies with price. In fact we have already seen at an intuitive level that the difference between monopolistic and competitive price movements depends on how the elasticity of demand varies with the price level. This is established formally by examining the consequences of a monopolistic seller equating the rate of change of net marginal revenue to the rate of interest. Incidentally, it may be worth mentioning that by extending the arguments of section 1, it is possible to establish that in the monopolistic case the impact of interest rate changes is similar to that in the competitive case.

1.3 DISCOVERIES AND DEMAND SHIFTS

The previous two sections have established that there are forceful reasons for a long-run upward trend in oil prices. What can interrupt such a trend?

We have already seen one factor, namely a rise in interest rates. There are others, perhaps more obvious. One is the discovery of extra reserves of oil. An unanticipated discovery of substantial extra reserves will clearly reduce the scarcity value of existing reserves. It will lead to a one-time drop in price, with the rising trend that was initially in place being followed again thereafter. Fig. 1.1 applies, except that in this case $r_1 = r_2$ and T is the date of discovery of the new reserves.

Another obvious possibility for causing a price drop is a drop in demand. The analyses of the previous sections assumed constant demand conditions. In fact an unanticipated drop in demand—a shift of the demand curve to the left—will have exactly the same impact as the discovery of extra reserves. The price will drop, and will thereafter resume its rising trend. This will be true whatever the reason for the drop in demand. This could stem from a policy-induced macroeconomic contraction, or from technological change allowing oil to be replaced in some of its uses (a possibility examined in detail in Chapter 2).

Overall, then, we have a picture of an upward long-run trend, interrupted or indeed temporarily reversed by demand drops, rises in interest rates,[8] and expansion in the reserve base. Of course, an increase in demand, a drop in interest rates, or a downward revision of the reserve base, will all have the opposite effect, reinforcing the upward trend.

1.4 INTEREST RATES AND RESOURCE PRICES

The previous two sections have expounded the theory that movements in the net price of oil will be influenced by interest rates, both in competitive and in monopolistic markets. The theory seems very reasonable, and hinges only on the assumption that producers of oil make some attempt to anticipate future prices and to produce and sell their oil in whatever period is most profitable. From this simple assumption follows a rich and non-trivial set of predictions about the effects of interest-rate levels and of changes in interest rates, and the effects of monopoly, on the movement of oil prices. However, ultimately a theory is judged not only by its simplicity and coherence, or by the insights that it generates, although these are very important issues: it is also judged by its predictive power. What then is the predictive power of the theories of section 1, 2, and 3?

The final verdict is still awaited on this issue. However, there is a growing body of empirical evidence[9] which substantiates the basic qualitative ideas of these theories. The price movements of a number of extractive resources (oil, copper, lead, zinc, and others) are statistically related to the behaviour of interest rates. Both in the short term, on a monthly basis, and over periods as long as a century, there has been found to be a correlation between changes in resource prices, and movements in interest rates. Models based on these relationships have shown predictive performances at least as good as any others.

In fact the statistical relationship between resource price movements and interest rates is considerably more complex than that implied by the simple theory, suggesting that, perhaps not surprisingly, a number of factors have been omitted from this theory. It appears that foremost amongst these factors are those relating to the manner in which traders in resource markets form their expectations about future prices. This is a key issue: the theory of the earlier sections is constructed on the assumption that traders, when deciding whether to sell now or at some future date, are sure what prices will be in the future. They are then in a position to balance the merits of present and future sales. In fact, of course, such certainty is seldom possible: instead traders use information about past price trends in an attempt to forecast future prices. Different ways of using this information will lead them to different purchase or sales decisions, and so will generate different price movements.

The empirical implementation of the theories we have reviewed therefore requires that the rather skeletal outlines given so far be augmented with information about how traders actually form their expectations, which of course makes the models more complex. Such information can be supplied, and with this elaboration the theories perform well. Unfortunately, there is also difficulty in obtaining data to test these theories: they relate movements in net price or net marginal revenue (net of marginal extraction cost) to interest rates. Data on price is available: data on marginal revenue and marginal extraction cost is not. Hence rather indirect methods must be used to test the theories. Although no clear conclusions have yet emerged, models based on the analysis of this chapter have systematically shown greater predictive power in markets for exhaustible resources than any other models.[10] The theory therefore has to be taken very seriously.

NOTES

1. This result was first stated formally by Harold Hotelling in a paper entitled 'The Economics of Exhaustible Resources', *Journal of Political Economy* 39 (1931), 137–75. Hotelling's article is technical, but there is an excellent non-technical summary of this and related results in Robert Solow, 'The Economics of Resources and the Resources of Economics', *American Economic Review*, Papers and Proceedings Issue 64 (1974). At an intermediate level, these issues are also covered in Partha Dasgupta and Geoffrey Heal, *Economic Theory and Exhaustible Resources* (Cambridge, 1979).
2. These arguments are set out in detail in the works by Solow and by Dasgupta and Heal, op. cit., n. 1.
3. One attempt to resolve it can be found in William Nordhaus, 'The Allocation of Energy Resources', *Brookings Papers on Economic Activity* 3 (1973). An extension of the arguments here is in the same author's *The efficient use of energy resources* (1979). See also chs. 11 and 15 of Dasgupta and Heal, op. cit.
4. This issue is explored further in Ch. 2.
5. See, for example, J. M. Griffin and D. J. Teece, *OPEC Behavior and World Oil Prices* (1982).

6. Dasgupta and Heal, op. cit., ch. 11.
7. The detailed arguments are given in Dasgupta and Heal, op. cit., ch. 11, and in Joseph Stiglitz, 'Monopoly and the Rate of Extraction of Exhaustible Resources', *American Economic Review*, vol. 6, no. 4 (1976), 655–61.
8. Macroeconomic contractions leading to downward shifts in demand are in fact often connected with upward movements in interest rates.
9. See Dasgupta and Heal, op. cit., ch. 15 n. 3. More details are given in two papers by Geoffrey Heal and Michael Barrow, 'The Relationship between Interest Rates and Metal Price Movements', *Review of Economic Studies*, xlv, 11 (1980), 161–81, and 'Empirical Investigation of the Long-Term Movements of Resource Prices: a Preliminary Report', *Economics Letters*, 7 (1981), 95–103. A review of the evidence is given by V. Kerry Smith in 'The Empirical Relevance of Hotelling's Model for Natural Resources', *Resources and Energy*, 3 (1981), 105–17. This earlier work is extended in Terence D. Agbeyegbe, 'Interest Rates and Metal Price Movements', *Journal of Environmental Economics and Management* (forthcoming). All of the above theories rely on tests of time-series data describing price movements over time. Merton H. Miller and Charles W. Upton, 'A Test of the Hotelling Valuation Principle', *Journal of Political Economy*, 93 (1985), 1–25, use pooled cross-section and time-series data on US oil and gas markets, and claim a substantial validation of the Hotelling theory outlined in the previous sections.
10. See references in footnote 9.

2

Technology and Oil Prices

In the previous chapter we had occasion to refer several times to issues which were essentially technological in nature. We referred, for example, to the possibility that the availability of substitutes for oil could place an upper limit on its long-run price, and also to the possibility that because of alternative fuel sources, the demand for oil becomes elastic at high prices. In this chapter we turn to these issues in more detail: they have in common a dependence on the technologies for oil use and for the production of substitutes for oil.

2.1 A BACKSTOP TECHNOLOGY

A concept which has proven useful in analysing the impact of technology and technological change on oil prices is that of a 'backstop technology'. This term refers to a technology that can be used to produce unlimited supplies of a perfect substitute for oil—at a price, of course. Fusion power, shale oil, and other synthetic fuels, have all at times been considered in this role. They all have the potential to produce power on a scale large enough to meet a very substantial share of the demands currently satisfied by oil, and to continue doing so over very long periods. Effectively they could act as substitutes for oil, though there is a great deal of uncertainty about the oil price at which it might become profitable to offer them in this role.

Suppose that the output of such a backstop technology were to be marketed competitively at a price equivalent to an oil price of P_B (in terms of the cost of a unit of useful energy). What would be the impact of this on oil prices? Clearly the price of oil would have to be less than or equal to P_B—for at prices above P_B, there would be no demand for oil. Buyers would switch to the backstop instead. All that remains is to see if the price of oil would in fact be less than, or instead equal to,

P_B. The argument of Chapter 1 section 1[1] can be used to show that in this case, as in the simple cases, the net price of oil does still rise, and rises at the interest rate. Obviously, these arguments can only apply until the market price equals the backstop price P_B. We therefore have a price path for oil as in Fig. 2.1: it rises just as before, until it reaches P_B.

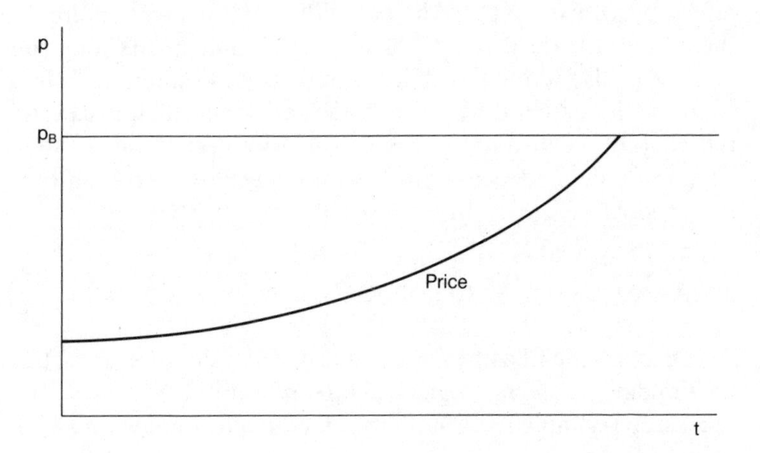

Fig. 2.1: The price path of oil when a backstop is available at P_B.

Why, it might be asked, will the price not just go immediately to the level P_B, or very slightly below this, and remain there as long as oil reserves are available? The answer is that although this might give a higher average price for oil sold, it would also delay the sale of the oil: higher prices mean lower demand and slower sales. So the possibility of higher total revenue has to be balanced against a delay in the realization of that revenue. Out of this trade-off comes the result in Fig. 2.1. (Note, however, that if oil supplies are monopolized, then there are conditions under which it pays to go straight to a price just less than P_B—see below.) So the net price, which is market price minus marginal extraction cost, rises at the interest rate until the market price equals the backstop price P_B. What can be said about the initial market price of oil, P_o, and the date T at which the market price of oil P_T first equals the backstop price P_B? In a competitive market, the initial price of oil P_o should be set by market forces so that if the net price rises at the interest rate, and T is the date at which the oil price P_T first equals the backstop price P_B on such a path, then the total reserves of oil are just depleted at this date T. So the bigger the reserves of oil, the lower is the initial price P_o and the larger is T, the date at which the oil price reaches the backstop level. There is an 'oil era' with the net price rising at the interest rate, oil less expensive than the backstop and all demand met by oil. Then the world shifts to a 'backstop era', with the price of oil substitutes constant at P_B and demand served by the backstop technology.

In such a world, what is the effect of a change in P_B, the price of the oil substitute produced by the backstop technology? Suppose for example that the price of the backstop's output falls from P_B to a value P_B' which is less than P_B. What happens to the price of oil? The answer is that the same conclusions as above apply, but now relative to the lower backstop price P_B' rather than to the old price P_B. So there is a new initial market price of oil, P_o', and a new date T′ at which the oil price P_T' equals the backstop price P_B'. Until T′, the net price rises from the initial value of (P_o' – MEC), and over this interval the entire stock of oil is depleted. It is easy to show that as the terminal price of oil in the oil era is now lower than before, so must be the initial price: reducing the cost of the backstop shifts down the whole price path of oil, as in Fig. 2.2.

Although this analysis is very simple, there are two important points that emerge from it, points with significant practical implications. One such point is that the price of oil will not be set by the price of the cheapest alternative to it, in this case the backstop price P_B. Just as some commentators (usually pro-OPEC) have argued that the competitive price of oil will be the price of the next best alternative, others (usually anti-OPEC) have argued that the competitive price of oil should equal its marginal extraction cost. Both are wrong: actually the competitive price of oil will range over time from near the former to the latter.

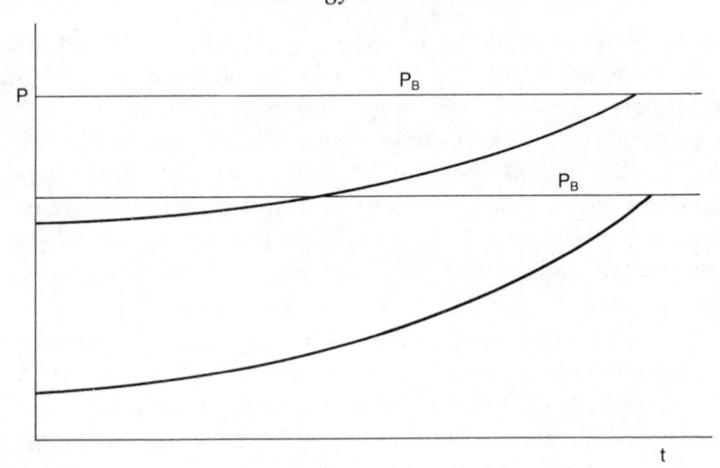

Fig. 2.2: The effect of a reduction in the price of the backstop.

Another very important point to emerge from our analysis is that even if the backstop technology is not currently being used, a reduction in its costs, and so in the price P_B at which its product could be sold, will lead to a lower price path for oil. So oil consumers could benefit right now from an improvement in the backstop technology, even if this technology were not to be used for another decade or more. This has interesting implications about the private and social returns to investment in backstop technologies: we shall return to them below.

How would these conclusions change if we were to consider a monopolistic rather than a competitive oil market? Very little, except that one further possibility is added. Under monopoly, we might see the price rising over time at such a rate that net marginal revenue grows at the interest rate, until the market price reaches the backstop price P_B: we might also see an interval during which the market price was set at, or slightly below, P_B. So a monopolist might raise the price gradually toward P_B, and then hold it just below the backstop price until his reserves were exhausted. This latter behaviour is an example of what is called 'limit pricing', i.e. setting the price at a level which limits the entry of competition into the market.

By setting the price just below P_B, the monopolist retains the oil market and prevents entry of the backstop technology. The relative lengths of the two types of interval—with price rising so that marginal revenue rises at the interest rate, and with a limit-pricing policy—will be very sensitive to demand conditions in the market.[2] Of course, the backdrop has a major effect on the price elasticity of demand: as the price of oil passes the backstop price P_B, demand for oil drops to zero, so that the elasticity becomes very great as the price of oil reaches the backstop price P_B.

2.2 A BACKSTOP TECHNOLOGY IN THE FUTURE

In the last section, we looked at the consequences for oil price movements of the existence of a backstop technology which could produce a perfect substitute for oil, and could do so right now if needed. Though a reasonable point to start the analysis, this is perhaps not the most convincing case. We do not yet have a backstop technology—but we hope that we might in the future.

All of the candidates for the role of backstop mentioned earlier—fusion, shale, synthetic fuels—are in the category of prospects rather than certainties. There is a great deal of uncertainty about the time horizons over which they might become available. In view of this, it is natural to ask what would be the effect of a backstop which is being developed, which will be available B years from now, but which will not be available at all in less than B years.

Again, this is not too difficult to work out. Over the next B years, the backstop cannot put a limit on prices. They can in principle certainly exceed P_B. However, from B years on, when the backstop is available, it places a cap on oil prices, i.e. P_B. In this case, under competitive conditions the market price of oil would rise over the next B years so that net price increases, as usual, at the interest rate. There are now two possibilities, shown in Fig. 2.3(*a*) and (*b*): which of these occurs depends on the balance between oil reserves and demand conditions.

Fig. 2.3(*a*) represents the case where reserves are in some sense small. Here the market price rises to a level in excess of P_B, along a path on which the total reserves will be be depleted at date B. At date B, oil is no longer available and the backstop takes over the market. Fig. 2.3(*b*) shows, in contrast, a situation where oil is relatively abundant. Now it is just as if the backstop were available immediately: the market price of oil rises until it reaches P_B at a date T later than B and after the backstop is available. At this date oil reserves are exhausted and the oil era ends. In the case of Fig. 2.3(*a*) a reduction in the backstop price P_B will have no effect on present oil prices: the link between the backstop price and the market price of oil is broken. In the case of Fig. 2.3(*b*), however, this link remains as in the previous section.

It is easy to understand what distinguishes these two cases. In Fig. 2.3(*a*), the date of availability of the backstop is sufficiently distant, and oil reserves are sufficiently small, that if the market price of oil were to follow a path reaching the backstop price P_B only at date B, then reserves would be completely exhausted prior to B. There would be a gap between the end of the oil era and the start of the backstop era, with nothing to meet demands for oil. It is the possibility of this gap which gives scope to oil producers to charge prices in excess of P_B. In

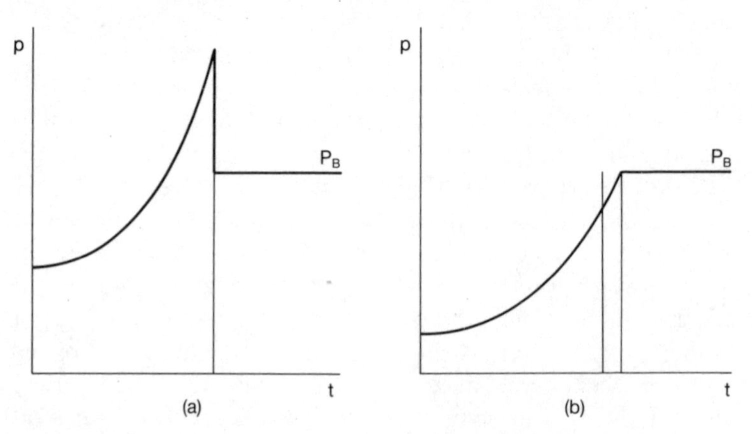

Fig. 2.3: A backstop available only B years ahead.

Fig. 2.3(*b*), however, a price path reaching the backstop price P_B at date B with the price rising at the interest rate, would leave unsold stocks of oil at date B. In turn this case backstop is available soon enough to influence the market price of oil.[3]

2.3 OTHER FORMS OF TECHNOLOGICAL CHANGE

Much research and development in the energy field is devoted to the development of possible backstop technologies—but not all. The rest goes to general forms of energy-saving, to the development of more energy-efficient equipment and structures. Usually these cost more than their predecessors, so that the effect is to give consumers a choice: they may pay a lower price for their energy-using equipment, and use relatively more energy, or pay more and use relatively less energy. This trade-off is clearly present in heating units, air-conditioning units, residential office structures, and in other types of equipment. In economic terms, the development of this choice which is driven by higher oil prices, makes the demand for energy (and for oil) more price-elastic. Its occurrence in response to higher prices of oil increases the scope that consumers have for lowering their demand in response to price rises.

How would this increase in the elasticity of demand affect price movements? In the competitive case, the whole profile of prices over time would be shifted down. In a monopolistic oil market, the price path would become flatter: in effect, what these developments do is make demand more responsible to price at high prices, and so move the system in the direction of Fig. 1.2 (case b) of Chapter 1.

2.4 PUBLIC POLICY TOWARD TECHNOLOGICAL CHANGE

There are many technologies that could be candidates for the role of backstop. One of the longer shots is fusion: in the nearer future, technologies might be developed to extract oil from unconventional oil deposits such as shale oil or tar sands. Reserves of these are extremely large and could in principle meet the world's energy needs for many years. At present, however, extraction technologies are still at a very preliminary stage of development. So these technologies fit most neatly into the framework of section 2.2—backstops available at some future date. However, we are certain of neither the date of their availability nor of the price of their output.

The important conclusion of the earlier sections was that a reduction in the backstop price could shift down the entire time-path of oil prices—even in the case of a backstop that is not yet available for use. This is the case shown in Fig. 2.3(*b*), in which current oil reserves are large enough to last until the backstop is available at a price less than the backstop price. In practical terms, a demonstration that oil from unconventional sources will definitely be available at, say, $40 per barrel, though perhaps not on a large scale for a further ten to fifteen years, could force down oil prices now and for the next ten to fifteen years. This in turn would lead to a transfer of wealth from oil producers to oil consumers. As the industrial countries are on balance consumers rather than producers of oil, they would stand to gain from such a demonstration. Note, however, that the company whose pilot plant established this point will not in general earn any returns on its invention for at least ten years. The industrial countries as a whole might gain more from such a discovery, than was gained by the firm making it. This establishes a strong case for a public subsidy to such research and development activities.

It is in fact noteworthy that the extent of such development activity is extremely sensitive to the current price of oil. In the late 1970s high oil prices induced a boom in the development of alternative oil sources, but the depressed oil prices of the 1980s have persuaded many companies to pull out of these projects. So if one of the effects of successful development in this field is to reduce the market price of oil, then it may clearly be self-limiting.

NOTES

1. A detailed statement of these arguments is given in Dasgupta and Heal, *Economic Theory and Exhaustible Resources*, ch. 6.
2. We could think of this as an extreme version of the case shown in Fig. 2 of Ch. 1, where the monopolistic price path is flatter than the competitive path because the elasticity of demand for oil rises as its price rises.
3. There is a formal discussion of these cases in Heal, 'Uncertainty and the Optimal Supply Policy for an Exhaustible Resource', in R. S. Pindyck (ed.), *Advances in the Economics of Energy and Resources*, vol. II (1979), 119–47. This article extends the analysis to cases where there is uncertainty about the date at which the backstop technology will become available—an important increase in realism, given that with all potential backstop technologies, major technological issues remain to be resolved.

3
The Demand for Oil

It is important to know how the use of oil changes and adapts with varying circumstances—and, in particular, to know how the use of oil changes in response to alterations in its price, and in the income levels of its consumers. The concepts of price elasticity of demand and of income elasticity of demand are used to formalize and quantify the changes in oil use as price and wealth vary. An analysis of these is the subject of this chapter.

3.1 PRICES AND DEMAND

For our purposes, it is responsiveness of oil demand to price that is most important. To understand the issues here, suppose that the price of oil has just increased very substantially: what follows? There are obviously a number of adjustments that a typical oil-user can make quickly and without undue effort to minimize the extra costs—turning down thermostats, wearing slightly warmer clothing, walking to a nearby store rather than driving, occasionally using public rather than private transportation, etc. So demand will fall after the price increase because of these and similar measures, but the fall will clearly not be great. Formally, in the short-run demand is relatively inelastic or unresponsive with respect to price.

Over a longer period, however, our representative consumer will make much bigger adjustments. The next car will be more fuel-efficient, more cotton will substitute for synthetic fabrics, the house will be insulated, and more efficient appliances will be installed. There will thus be further falls in demand in response to the original price rise, but they may be completed only many years after the price increase, so that in the long run, demand will be much more elastic with respect to price than in the short run.

What is true of individual demand for oil, is also true of industrial demand. A price increase brings in its wake short-run responses, which are rather limited, and then long-run responses, which are much greater. For example, airlines responded immediately to higher fuel prices by flying their planes slightly more slowly, and by rescheduling to consolidate flights. In the longer run, they use aircraft which are more fuel-efficient: this has a much greater impact on their fuel demands.

This phenomenon of a long-run response to price changes which is much greater than the short-run response, is sufficiently general and important that it is useful to illustrate it by reference to a particular case. Consider how the demand for petrol as a fuel in motor vehicles changes in response to a major increase in the real price of petrol.

Vehicle owners would obviously search for ways of minimizing the resulting rise in operating costs. They could increase the fuel-efficiency of their existing vehicles by driving them more slowly (fuel consumption varies with the square of a vehicle's speed), and by maintaining them in better tune. They could also abandon using motor vehicles for very short journeys, on which fuel consumption per mile is very high, and walk or bicycle. And on certain routes they might turn to more fuel-efficient public transport. All of these are measures which could be introduced immediately after a price increase, and which would have an immediate effect on demand. It is difficult to estimate how big a demand reduction they could produce: probably it would be at most of the order of 10 to 20 per cent.

Figs. 3.1 and 3.2 develop this point further. Fig. 3.1 shows that from 1973 to 1985, the fuel efficiency of US passenger cars rose by 30 per cent, having fallen by about 5 per cent from 1967 to 1973. Miles travelled per car also showed a slight downward trend after 1973, having been rising slowly prior to 1973. Fig. 3.2 shows that new car fuel efficiency rose much faster post-1973 than did 'fleet' or average fuel efficiency.

In the longer run, vehicles wear out and are replaced: this provides their owners with a much more important opportunity to adjust patterns of fuel to use to the new petrol price. There are two stages in this process. In the first, vehicle manufacturers

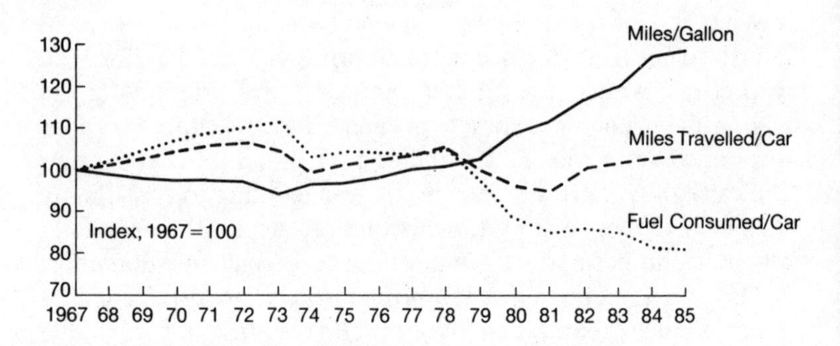

Fig. 3.1: US passenger car efficiency.

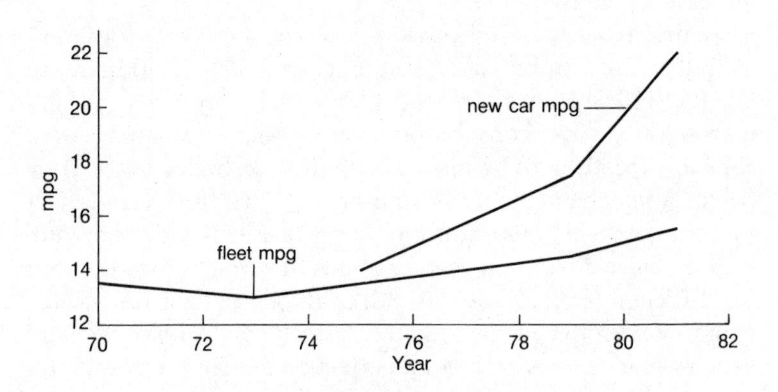

Fig. 3.2: On-road efficiency of US passenger cars.
Source: DOE, *Energy Conservation Indicators Annual Report* (1982), p. 47.

anticipate that high oil prices will shift patterns of demand towards more fuel-efficient vehicles. They therefore begin the research and development necessary for the production of such vehicles. This may require experimenting with different ignition systems or combustion systems, experimenting with different metals to achieve a lighter engine, and experimenting with body designs to reduce wind resistance. Obviously, the completion of this experimental stage could easily require five years. Next, the new designs have to be put into production, requiring the design and installation of new machine tools and assembly lines. This will again require several years: all in all, it could be eight or nine years after the initial oil price increase before one sees its fruits in terms of more fuel-efficient vehicles in the showrooms.[1]

It is only at this stage that vehicle owners can begin to transform their vehicle stocks by moving to the more fuel-efficient designs. They will not, of course, change immediately: they will purchase newer and more efficient vehicles as the old ones need replacing—though the increase in operating costs from higher fuel prices may increase the rate of replacement. Suppose that the average life of a motor vehicle is ten years, so that the entire vehicle stock is replaced on average every ten years: then it will be ten years after the introduction of the new designs before the entire vehicle stock embodies the more efficient technology, making a total delay of the order of eighteen years from the initial price increase. Note that though this process is very gradual, it can lead to major changes in fuel demand: by 1983, a decade after the first major oil price rise, the average fuel-efficiency of new automobiles sold in the USA was 85 per cent higher than the fuel-efficiency of new automobiles sold in 1973, implying that with constant driving patterns demand for fuel for private automobiles would decline by 85 per cent relative to 1973 once the entire stock of private vehicles embodied the new technology.

A very similar analysis holds for other sources of demand for oil, though there the lags in response may be even longer still. Consider, for example, the use of oil as a domestic heating fuel. In the first few years after an oil price rise, demand is reduced by turning down thermostats, and by elementary insulation measures. In the long run, heating systems are replaced, and consumers switch to other fuels, or to newer and more efficient oil-burning furnaces. (From 1973 to 1983 the thermal efficiency of best-practice oil-burning domestic furnaces rose from around 50 per cent to over 80 per cent.) New houses enter the housing stock, with very much better insulation and much lower heat losses than could ever be achieved by improvements to the existing housing stock. Obviously, we are talking here of a process that may not be completed for many decades, as the turnover of the furnace and housing stocks is a very gradual process. In the industrial field, similar considerations apply to energy consumption in office buildings: in power stations, or in energy-intensive industries such as steel, major changes in energy demand patterns can only come as existing capital equipment is replaced or substantially renovated, an expensive and time-consuming process.

3.2 INCOMES AND DEMAND

Matters are rather simpler when we discuss the response of oil demand to changes in income levels. Here there does not seems to be the great divide between immediate and long-run effects: all the consequences of an income change appear within a year or so. The word *income* is being used here to refer to national income, which is essentially the aggregate of all incomes, from work or from capital ownership, received by the citizens of a country. Variations in national income in this sense are closely correlated with variations in industrial activity, as well as with variations in individuals' wealth.

Energy use in the industrial sector can be broken into two categories—overhead use, and process use. Overhead use includes use in offices, use in factory lighting and heating, use in communication, and use in research and development. Process use denotes the use of energy in production processes—to heat steel furnaces, drive assembly lines power-machine tools, and transport raw materials. It is to be expected that the process uses of energy will vary with the level of industrial activity, and hence the level of national income, and that overhead use will not. Hence in aggregate we expect industrial energy consumption to vary with the level of national income, but to do so less than proportionally.[2]

A similar conclusion applies to the private sector's use of energy: each household has certain minimal demands for energy for heating, lighting and cooking, which it will continue to satisfy even under very adverse circumstances. Additional demands for energy will vary with income levels, as higher incomes lead for example to more travel, or to the purchase of larger cars or of more electronic equipment. So we would again expect a variation of demand in response to income, but one which is less than proportional.

3.3 THE STATISTICAL EVIDENCE

The responses of demand to price and income changes are, as already mentioned, important matters. They have technical names—price elasticity of demand, and income elasticity of demand. A price elasticity of demand of for example −0.5 means that a 1 per cent increase in the price of a good leads to a fall of 0.5 per cent in demand for that good: an elasticity of −3 would imply that a 1 per cent rise in price produces a 3 per cent drop in demand. An income elasticity of demand of 0.8 implies that a 1 per cent rise in income produced a rise in demand of 0.8 per cent, etc. For simplicity we will abbreviate the phrases price elasticity of demand and income elasticity of demand to P-elasticity and I-elasticity respectively.

Our earlier discussions imply that the P-elasticity for oil will depend on the response period that we consider. If we look at the response of demand within one or two years of a price change, the P-elasticity will be a small number: however, if we look at the eventual long-run response, it will be much greater.

A tremendous amount of research effort has been devoted to producing estimates of the P-elasticity and I-elasticity for oil. For a number of technical statistical reasons, this is a very difficult task: there is disagreement about the best estimates, and the available results can only be regarded as tentative. In spite of this, they are enlightening, and help to provide insights into the dynamics of the oil market. In fact the estimates that we have relate not to the demand for crude oil, which is an intermediate product used only to produce other goods, but to the demand for energy by all users, industrial and private. In interpreting these estimates, we therefore have to bear two facts in mind. One is that estimates of the P-elasticity and I-elasticity for energy may give a misleading picture of how demand for oil changes in response to price, if the price of oil varies and other energy prices remain constant. They show how total energy demand changes as the average price of energy varies: if in the process some types of energy experience bigger price increases than others, then substitution between fuels could cause much larger demand responses for these than would be predicted by the overall elasticity. In fact most energy prices have moved roughly in parallel in the last decade, so that this will not be a major problem in practice.

A second fact to be borne in mind in making deductions about the demand for oil from the P-elasticity for energy, is that a 10 per cent rise in the price of crude oil does not necessarily imply a 10 per cent rise in the prices to their consumers of oil-based products. These prices depend on the price of crude oil, the costs of transportation and processing, the profits of the oil companies, and taxes levied by the governments of the consuming countries. Table 3.1 shows how the cost of a gallon of gasoline breaks down between these headings: as the cost of crude oil is only about 50 per cent of the total price to the consumer, a 10 per cent rise in the price of crude will raise the retail price of gasoline by only about 5 per cent. The drop in demand will therefore be 5 per cent times the P-elasticity, and not 10 per cent times the P-elasticity.

With all these qualifications clearly set out, we can now turn to the statistical estimates. On the income elasticity of demand for energy, there is general agreement: the best estimates of this are about 0.8. So as income rises or falls, energy consumption rises or falls, but by a smaller proportion. This is entirely consistent with the earlier observation that some fraction of energy use is in the 'overhead' category, or its equivalent in the private sector, and so will be unaffected by changes in income.

There is less general agreement on the price elasticity of demand for energy. Estimates of the short-run P-elasticity vary between −0.2 and −0.5. Good estimates of the long-run P-elasticity are very scarce—primarily for the rather obvious

Table 3.1: Breakdown of cost of US gallon of gasoline (cents per gallon)

	A	B		C	
Year	Service station price (incl. tax)	State and Federal Tax	State and Federal Tax as of service station price (%)	Cost of crude per gallon	Cost of crude as a percentage of service price
1974	52.41	12.00	22.9	22	41.20
1975	57.22	11.77	20.6	25	43.19
1976	59.47	12.03	20.2	26	43.60
1977	63.07	12.37	19.6	28	45.15
1978	65.71	12.62	19.2	30	45.15
1979	87.79	13.46	15.3	42	48.06
1980	121.72	14.37	11.8	67	54.91
1981	131.10	n.a.	—	84	64.00
1982	122.20	n.a.	—	76	62.10
1983	115.70	20.34	17.6	69	59.66
1984	121.20	20.84	17.2	68	56.24

Source: *Basic Petroleum Handbook* (1986).

reason that the full long-run impact of the 1973 energy price rise is only now starting to be felt so that it is hard to measure. One can measure what has happened so far and extrapolate, but this is a risky business. With this caveat in mind, it is reasonable to say that there is a convergence of opinion that the long-run price elasticity of demand for energy is about −1.25. This makes it more than twice as high as the average estimate of the short-run elasticity.[3]

3.4 PATTERNS OF FACTOR DEMAND

We have discussed so far the price and income elasticities of demand for energy. In the field of energy demand analysis, these are the most important parameters, as they tell us how energy use will respond, in aggregate, to price and income changes.

There is another set of parameters which are also of great concern: These are the *substitution elasticities* between oil and other factors of production, particularly capital. The issue to be addressed here is: are energy and capital equipment substitutes for each other, so that an energy price increase leads to an increase in demand for capital equipment, or are they complements, so that an energy price rise leads to less use of energy and less use of capital as well? The importance of the issue is that the answer will tell us how demands for capital equipment are likely to respond to energy price increases, and will also give us insights into the process by which the economy adjusts to changes in energy prices.

There are two mechanisms by which the balance between energy and capital use can respond to a change in the relative prices of these two factors: one we shall term the microeconomic mechanism, and the other the general equilibrium mechanism.

The microeconomic mechanism is rather obvious, and is the response pattern of which we have already seen several examples. Higher energy prices lead to an increased demand for items of capital equipment associated with energy conservation, which may be insulation devices, more efficient furnaces, more efficient vehicles, more efficient aero engines, etc. In all of these cases, capital equipment is being used at a microeconomic level to replace energy, establishing a relationship of substitutability between capital and energy. There are other simple microeconomic examples which illustrate a relationship of complementarity between energy and capital—less energy consumption will lead, for example, to reduced demands for fuel tanks, pipelines, and other items directly linked to energy use.

The general equilibrium mechanism is more complex, and involves changes in the overall configuration of the economy (its 'general equilibrium') in response to an energy price change. For example, if energy prices rise, then the prices of products and services to which energy is a major input will also rise. Transportation and steel will become more expensive relative to food and clothing. This change in the relative prices of goods and services will lead to a change in the proportions in which they are consumed, with demand shifting away from those which are energy-intensive and whose price has risen, towards those which are less energy-dependent. This change in the overall pattern of consumption in society will lead to a drop in demand for energy and an increase in demand for the types of capital equipment used to produce the goods and services which are not energy-intensive. Through this mechanism, an energy price increase has led to an increase in demand for certain types of capital, and hence to an appearance in the aggregate of substitutability between energy and capital.

One general equilibrium effect of an energy price increase is a change in the pattern of final demand. Another may be a macroeconomic contraction because of the deflationary impact of higher energy prices (an issue discussed at length in Chapter 6). To the extent that the overall level of activity in the economy falls because of an increase in energy prices, this will lead to a drop in demand for capital equipment, and hence at the aggregate level to an appearance of complementarity between energy and capital equipment.

So the general equilibrium mechanism by which the balance between capital and energy use may respond to an energy price change can produce complementarity or substitutability between energy and capital. A detailed analysis is needed to determine the circumstances under which each will occur.[4] What is interesting about this mechanism is that it may lead a particular economy, with a fixed set of production technologies, to show substitutability from some initial price levels and demand patterns, and complementarity from others. This in turn can explain the puzzling empirical fact that, although Western Europe and the USA certainly have the same energy-using technologies, Western Europe displays substitutability between energy and capital whereas the USA displays complementarity.[5] According to the general equilibrium approach, this could be due to the very different energy price levels and demand patterns in these regions.

NOTES

1. During this process countries whose vehicles were initially relatively fuel-efficient will gain a competitive advantage in international markets.
2. It is worth noting that in the service sector, which is gaining importance in the advanced industrial economies, energy use is likely to be even less directly linked to output. So the development of this sector may reduce the income elasticity of energy demand.
3. For a more detailed reading on this subject, see R. S. Pindyck, *The Structure of World Energy Demand* (1979); W. D. Nordhaus (ed)., *International Studies of the Demand for Energy* (North Holland, 1977); and G. S. Maddala, W. S. Chern and G. S. Gill, *Econometric Studies in Energy Demand and Supply* (1978). See also Heals's review of books on energy demand in *Economica* (1979), 322–3. The most recent study is W. W. Hogan, 'Patterns of Energy Use Revisited', discussion paper (Kennedy School, Harvard University, 1988).
4. Such a detailed analysis is given in Chichilnisky and Heal, 'Energy-Capital Substitution: A General Equilibrium Analysis', collaborative paper, International Institute for Applied Systems Analysis (Laxenburg, Austria, 1983).
5. See, for example, E. R. Berndt and D. W. Wood (1975), 'Technology, Prices and the Derived Demand for Energy', *Review of Economics and Statistics*, 56, 259–68; E. R. Berndt and D. O. Wood, 'Engineering and Econometric Interpretations of Energy-Capital Complementarity', *American Economic Review*, 69 (1979), 342–54; M. Denny, M. A. Fuss, and L. Waverman, 'The Substitution Possibilities for Energy: Evidence from US and Canadian Manufacturing Industries', Working Paper No. 8013 (Institute for Policy Analysis, Toronto 1980); and J. M. Griffin and P. R. Gregory, 'An Intercountry Translog Model of Energy Substitution Responses, *American Economic Review*, 66 (1976), 845–57.

4

The Role of OPEC

At least from 1973 to 1980, it was widely taken as self-evident that OPEC exerted substantial monopoly power in the world oil market. The evidence advanced in support of this view was varied. Sometimes it was the rising trend of oil prices, sometimes the great excess of market price over extraction costs.

Chapter 1 taught us to beware of these arguments: none of them are convincing. In fact Chapter 1 showed that it would be very hard to establish a clear-cut case for OPEC gaining monopoly power in 1973. Differences between monopoly and competitive pricing are very sensitive to the demand conditions prevailing in the oil market. Perhaps the most plausible case is that of Fig. 1.2 (case b), where the price elasticity of demand for oil rises as its price rises, so that the monopoly price path starts higher than the competitive, but rises more slowly. In this case transition from competition to monopoly has the effect shown in Fig. 1.5: a once-and-for-all price rise, followed by a flatter price profile. This could be said to explain what happened between 1970 and 1978—see Fig. 4.1. So the theory certainly does not rule out interpreting the data as indicating that OPEC monopolized the world oil market in 1973—but nor does it rule out other inferences. In this chapter, we explore the inferences that can be made about the nature of competition in the oil market, and about the role of OPEC.

4.1 WHO SETS OIL PRICES?

So far we have used the word 'monopoly' without definition, confident that most readers will have no difficulty attaching some kind of meaning to it. Now we have to look at this issue more carefully: what, precisely, do we mean by monopoly

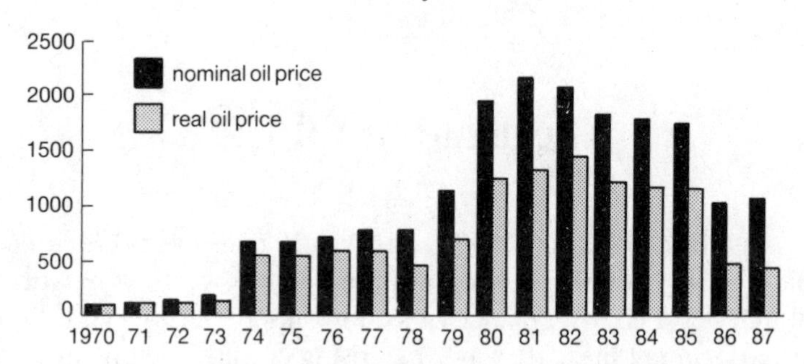

Fig. 4.1(a): Nominal and real oil prices: real prices obtained by deflating using OECD GDP deflator.

Sources: *UN Monthly Bulletin of Statistics* and *OECD Economic Outlook.*

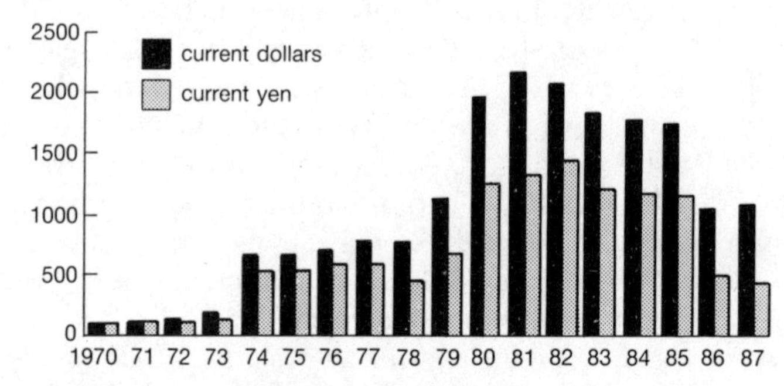

Fig. 4.1(b): Index numbers of current oil prices in dollars and yen.

Source: *UN Monthly Bulletin of Statistics.*

power? The essence of any answer must lie in the ability to influence the market price, rather than simply having to take it as given. There are problems in applying this definition to oil. One is that there are several ways of measuring the price, and it is not immediately clear which one should be selected. A second problem is that the concept of monopoly is essentially what economic theorists call a 'partial equilibrium' concept: 'partial equilibrium' analysis involves the study of one market at a time. The concept of monopoly has no analogue in a 'general equilibrium' context, i.e. in the context of an analysis which looks at the economy as a whole. The impact of oil prices changes and the power to effect such changes are, however, matters which have to be evaluated in a general equilibrium framework. We will look next at the issue of price measurement, and then turn to the question of what one means by monopoly power in a general equilibrium context.[1]

Fig. 4.1 (*a*), (*b*) shows the movement of several different measures of the price of crude oil. The first is the price of crude oil denominated in current US$ per barrel, the nominal price of oil. The second is the price of crude oil measured in constant US$, and the third is the price of oil in Japanese yen. As the rate of exchange between dollars and yen has varied substantially the first and third series give rather different time profiles for the price of oil. It is the first and second that matter to the Americans, and the third to the Japanese.

It is already clear that 'setting the price of oil' is not a phrase with an unambiguous meaning. In fact *none* of these three prices is the price which matters to OPEC. OPEC members export oil in order to import various types of goods and services: to date, these have been primarily military and industrial goods, and construction services. So the price of oil that matters to them, the measure of the purchasing power of their oil, is the price of oil relative to these imports. This price will be referred to as OPEC's *terms of trade*. When economics textbooks define monopoly power in terms of the ability to set prices, it is to prices in this terms of trade sense that they refer: they work in a framework where all prices other than the output price are fixed, so that a change in the output price automatically translates into an equivalent change in the price of the output relative to any other good or bundle of goods.

Terms of trade in this sense are not easy to measure, because it is not clear what is the appropriate deflator. Fig. 4.2 shows two possible deflators: the first is an index of OECD export prices to the Middle East. The second is the OPEC Import Price Index. Although these two measures of OPEC's import prices move rather differently, both rose very substantially over the period 1973–80, their average increase being nearly 300 per cent.

We can now return to our original concern—OPEC's ability

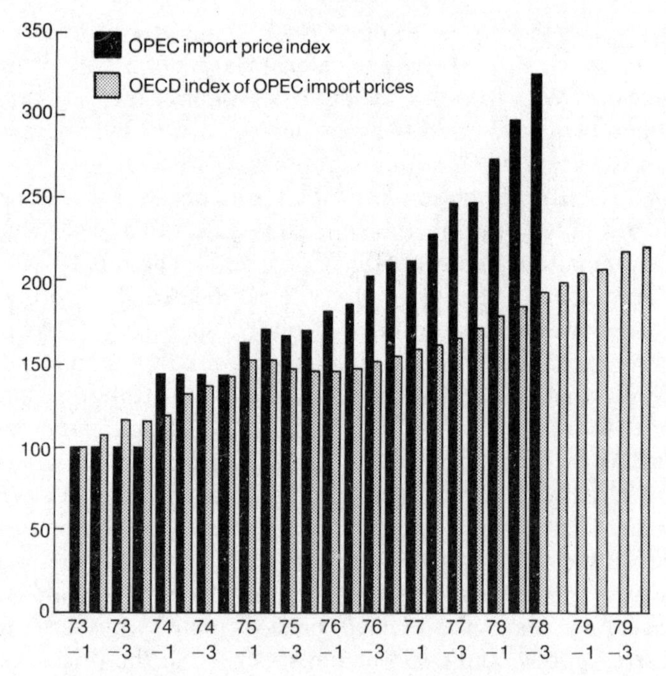

Fig. 4.2: OPEC import prices—by OPEC and OECD.

to set, and in particular raise, the price of oil. The price which OPEC might control, at least in some degree, is the first price that we considered, the nominal US$ price of crude oil. As the figures above show, this moves quite differently from the price that matters to OPEC, the price of oil relative to the prices of OPEC's imports. Its ability to set the former certainly would not imply an ability to set the latter. The terms of trade are a ratio: OPEC plays a role in setting the numerator, but the denominator is determined in the countries with whom OPEC's members trade. In principle, these countries could collectively respond to a rise in current US$ oil prices by an equal proportional rise in their export prices to OPEC, keeping OPEC's terms of trade constant. They have not done this—but their export prices have risen much faster than their domestic prices, and in particular have risen enough to make the rise in OPEC's terms of trade much less than the rise in current dollar oil prices. The mechanism by which this occurred is not entirely clear. In part it may be a consequence of inflation in the OECD generated by higher oil prices, though as we shall see in Chapter 6 the extent of oil-price induced inflation has not been very great. In part it is also attributable to exchange rate movements, but this still leaves a substantial component unexplained.

In summary, we can think of the real price of crude oil to OPEC as being set jointly by OPEC, by non-OPEC oil producers, and by OPEC's trading partners. It is a ratio: OPEC influences the numerator, and its partners set the denominator. There is of course an asymmetry in the roles here. OPEC members clearly do get together to exert a conscious influence over the numerator: there is no equivalent act on the part of its trading partners. The OECD countries make no collective decision on the level of their export prices to OPEC. Perhaps it is worth remarking, however, that they could do: between them they control a very substantial fraction of world trade in the goods and services that OPEC imports. In practice the OECD's export prices to OPEC, although not consciously chosen in response to oil prices, are influenced by many of the macro and exchange-rate policy variables chosen by the OECD governments, and have moved in such a way as to cancel much of the rise in nominal oil prices. In fact the macroeconomic policies of the oil-consuming countries influence OPEC's terms of trade not only via their effect on export prices and the denominator: by shifting the demand curve for oil they affect the current dollar price resulting from any given OPEC production policy. And in previous chapters we have seen other mechanisms through which the industrial countries can influence oil price—through interest policy (an aspect of general macro policy), and through policies towards technological development. Overall, the picture is far from one of OPEC as a price setter: oil prices result from a price-setting game played out between OPEC, non-OPEC oil producers, and their customers.[2]

An important point which emerges from this discussion is that the OPEC cartel, even if fully cohesive internally, cannot be considered to be a monopolist in the traditional partial equilibrium text-book sense of a seller who is able to set the real price of his output so as to maximize profits, taking as given the prices of other goods and services, and in particular taking as given the prices of the goods and services which he purchases from the rest of the world.

This is the standard picture of a monopolist—someone who buys inputs and sells outputs, and is able to choose the price of the outputs optimally knowing that this choice has no impact on input prices. This is the essence of the partial equilibrium picture: we just look at changes in one market, without considering that these may lead to alterations in other markets which can feed back to the market which we are studying. To study this full range of feedbacks is the object of the general equilibrium approach to economics.

This is an illustration of a general point. A change in the price of a good as important as oil is bound to have an impact on a wide range of other prices, which in turn will feed back to the oil market. For example, a change in the nominal price of crude oil will affect the prices of goods and services imported by oil exporters, and so have a second-round effect on their terms of trade.

Because of this, a general rather than partial equilibrium approach must be taken. Here there is no clear analogue to the naïve partial equilibrium concept of monopoly, as in this framework a change in one price will in principle affect all other prices through changing production costs. The rational price setter will take this into account. Such changes will also in principle lead other price setters to reconsider their pricing policies, and again in a general equilibrium context the rational price setter will need to anticipate this. These issues are relevant to oil: they are also relevant to the pricing of other important commodities (such as copper or bauxite) and to prices such as the price of capital (the rate of interest). The implication is that in any such market, a seller will have to consider the effect of changes in his price on other prices, and also on the actions of other participants in the market. We are inevitably forced to analyse the interplay between all of the participants in the market, as outlined above for the oil market. This general equilibrium approach is pursued in the analysis of chapters 6, 7, and 8 below, where we always emphasize the importance of looking at interactions between the oil market and other markets.[3]

4.2 OPEC'S AIMS

OPEC's role in the oil market can only be understood if we have an idea of OPEC's aims. Like many other issues we shall turn to, this is not straightforward. It is not straightforward partly because no economic agent's aims are ever straightforward, and partly because there are several groups in OPEC, each with apparently different objectives.

Nominally OPEC admits to several aims—the maximum rate of economic development of its member countries, the preservation of the world's reserves of hydrocarbon fuels, and the maintenance of economic balance in the oil-consuming countries. The first of these—maximizing the rate of development of its member countries—presumably translates into a pricing policy which maximizes the wealth attainable from members' oil reserves. It is at this point that a complication arises: as OPEC's members have oil reserves of very different sizes and geological characteristics, and have economies with very different characteristics, a policy which maximizes the wealth attainable from member A's reserves may be quite distinct from that which achieves the same end for member B. Put simply, the different reserve positions and economic structures of its members give them very different economic interests in the world oil market.

One has to understand this point in order to appreciate the dynamics of OPEC's pricing policy. Table 4.1 shows the economic characteristics of various OPEC members, giving their reserves both absolutely and in terms of the number of years for which they can produce at 1985 levels, and also population and oil reserves per head. There is a group of

Table 4.1: OPEC countries: 1985 population, reserves, and per capita revenues.

	Population (million)	Proven reserves (MMB)	Output (1000 B/D)	Years at 1985 output rate	Revenue from oil exports (US$M)	Revenue per inhabitant (US$)
Group I						
Saudi Arabia	11.52	171,490	3,175.00	147.98	27,500	2,387
Libya	3.77	21,300	1,023.70	57.01	10,924	2,897
Kuwait	1.71	92,464	963.30	262.98	9,690	5,667
Qatar	0.30	4,500	290.10	42.50	3,225	10,750
UAE	1.33	32,990	1,056.80	85.53	12,492	9,393
SUBTOTAL	18.63	332,744	6,508.90	140.06	63,831	
Avg. revenue per inhabitant				3,426		
Share of total (%)	4.90	65.28	43.02		49.70	
Group II						
Iran	44.75	59,000	2,192.30	73.73	12,881	288
Venezuela	17.32	27,200	1,681.00	44.33	10.347	597
Iraq	15.90	65,000	1,404.40	126.80	12,459	784
Algeria	21.60	8,820	672.40	35.94	8,178	379
SUBTOTAL	99.57	160,020	5,950.10	73.68	43,865	

Avg. revenue per inhabitant						441
Share of total (%)	26.10	31.39	39.33		34.20	
Group III						
Nigeria	99.59	16,600	1,491.10	30.50	12,185	122
Indonesia	163.39	8,500	1,178.20	19.77	8,557	52
SUBTOTAL	262.98	25,100	2,669.30	25.76	20,742	
Avg. revenue per inhabitant						79
Share of total (%)	68.90	4.92	17.64		16.10	

Excludes Ecuador and Gabon.
Source: *OPEC Annual Statistical Bulletin* (1985).

members with high reserves and high revenues per capita, mainly the Gulf States, Saudi Arabia, Kuwait, Qatar, and United Arab Emirates. At the other extreme are members—Indonesia and Algeria—who have much lower reserves and lower revenues per capita. Members of the first group, in addition to high oil reserves, have little else as a source of wealth. They have no substantial agricultural or industrial sectors, no commodity sectors, and no tourist industries. They are desert countries, living off the revenues from oil sales. The low-reserve countries typically have more diversified economies: they have substantial domestic agricultural sectors, they export commodities (Nigeria exports cocoa, Indonesia wood), they have a light industrial base, and in some cases (Algeria) have the potential to earn revenue from tourism.

These two groups can look forward to very different economic futures. The high-reserve per capita group will continue to depend primarily on oil for their income for many decades: they will still be living mainly off oil revenues well into the next century. The second group, however, will run out of oil reserves well before then. By the turn of the century they will be marginal oil producers, and other sectors of their economies will have to expand as they diversify out of oil. The high-reserve group therefore have a long-term interest in the world oil market: they expect to continue their dependence on it effectively indefinitely. The others can see themselves and the oil market parting company—if not soon, at least in the foreseeable future.

These differences naturally give rise to rather different positions on oil pricing policy. The long- and short-run consequences of a given pricing policy can be very different, because of the differences between the long-run and short-run price elasticities of demand for oil. Over a period of up to five years, or possibly even more, a big price rise may lead to little reduction in demand, and so to a big increase in revenues. In the longer run, however, the processes of substitution of other fuels for oil, of energy conservation, and of changes in demand patterns, processes which were initiated by the price increase, will erode the demand for oil and reduce the sellers' revenues.

Indeed, high prices maintained over a long period will provide a stimulus to the development and improvement of backstop technologies, with a resulting loss of revenues to oil producers. In this case, we are talking of a long-run process—probably fifteen years at least are needed for major improvements in such technologies.

It is clear, then, that a country that expects to be in the oil market in a major way for at least another half century, such as Saudi Arabia, will have a different view on pricing policy from say Indonesia or Algeria, which could be substantially out of the oil business within twenty years. If the price is raised substantially now, Saudi Arabia will still be selling oil when the long-run consequences begin to be felt. Low-reserve countries, however, will see much less of the long-run impact. And as the short-run impacts of a much higher price would be favourable to oil producers, and the long-run impacts possibly unfavourable, it is natural that countries in the low-reserve group will be much more favourably inclined towards high prices than those in the high-reserve group. In many markets one finds this distinction between those sellers who plan to stay in the market, and are concerned not to 'spoil' it in any sense, and those whose association is less durable and whose policies have a shorter time-horizon.

Commentators often divide OPEC members into 'hawks' and 'doves', depending on their position on pricing policy: those who favour high prices such as the Iranians are 'hawks', whereas the Saudis and others in favour of lower prices are 'doves'. Membership of these groups is frequently explained in political terms: radical regimes are 'hawks', and conservative pro-Western regimes are 'doves'. This explanation does not seem to fit the facts fully: Iran, for example, was a price 'hawk' even under the Shah, who was certainly conservative and pro-Western. Venezuela has also been hawkish on prices, though by no means particularly radical or anti-Western: the same is true of Algeria. This is not to deny that political considerations may influence OPEC members' positions on prices, but simply to say that there are straightforward matters of economic self-interest that seem to have a lot of explanatory power.

The behaviour of Saudi Arabia within OPEC illustrates very clearly the importance it places on maintaining 'moderate' oil prices in the long-run. Recall that, as shown in Fig. 4.1, the nominal price of crude oil was increased sharply in 1978 and 1979. There was disagreement within OPEC about the appropriateness of the new price level, with the Saudis openly of the opinion that $34 per barrel was too high. As demand for OPEC oil started to decline for other reasons about 1980, the price level of $34 proved hard to maintain, and most OPEC members cut back production: Fig. 4.3 shows the levels of crude oil exports of Saudi Arabia, and of OPEC as a whole.

The Saudis were in fact actually increasing exports as other OPEC members were cutting back their exports, and maintained a high export level until early 1982. Then when the official price of crude was reduced, the Saudis cut exports back very sharply indeed in order to help maintain the price of which they approved. This behaviour can only be interpreted as an attempt to use their considerable productive capacity to enforce their own preferred pricing policy when that selected by OPEC differed from it. As long as official OPEC prices were higher than Saudi Arabia wished, they maintained high

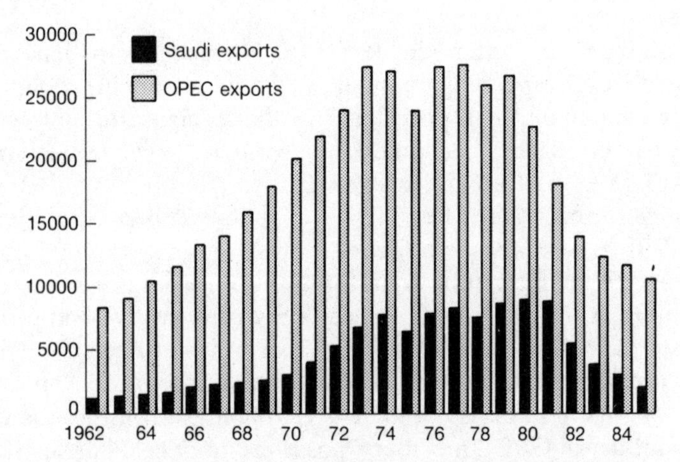

Fig. 4.3: Saudi exports and total OPEC exports (000 bbl/day).
Source: *OPEC Annual Statistical Bulletin* (various issues).

production levels—even when others were cutting back—to put downward pressure on prices. As soon as the official OPEC position changed, in conformity with their wishes, Saudi Arabia cut back output and backed the official price.

In fact this incident illustrates well the important role played by Saudi Arabia in OPEC. Their power derives from two factors. One is their large production capacity—at 11 million barrels per day, by far the largest in OPEC. The second is their capacity to change their output rapidly—as exemplified by their cutback in early 1982 to as little as 2.5 million barrels per day, in the process taking almost 10 million barrels per day of oil off the world market, almost one-third of the total volume of internationally traded oil. They can cut back in this way because of their wealth: the consequences of foregoing export revenue are less serious for them than for countries such as Iran or Nigeria which are short of foreign exchange. This ability to expand or contract output at short notice by amounts of up to one-third of the volume of traded oil, is one which gives Saudi Arabia a unique and pivotal position in the process of price determination.

4.3 NON-OPEC PRODUCERS

In fact, Saudi Arabia is not the biggest oil-producer in the world. As Table 4.2(*a*),(*b*) shows the world's biggest producer is the Soviet Union, followed by the USA and Saudi Arabia. In 1981 OPEC as a whole produced of the order of 40 per cent of world oil output (Fig. 4.4 shows how this has varied over time) and about 70 per cent of internationally traded oil (Fig. 4.5). Fig. 4.6 shows that OPEC members have been above 65 per cent of the world's proven reserves of oil. Table 4.2(*b*) shows that non-OPEC production of oil has risen rapidly since 1973, a result of the profitability of previously unprofitable fields at higher prices of oil.

It is clear from this that OPEC is by no means a complete monopolist in the world oil market, though it is a major actor. The crucial difference between OPEC producers and non-

Table 4.2(*a*): OPEC oil production.

Thousand barrels daily	1973	1979	1985	1986	1987
Saudi Arabia	7690	9840	3750	5540	4650
Iran	5900	3200	2250	1930	2350
Iraq	2020	3490	1460	1680	2090
Venezuela	3460	2430	1740	1880	1890
UAE	1530	1850	1350	1540	1700
Kuwait	3090	2630	1070	1440	1390
Nigeria	2050	2310	1510	1470	1330
Indonesia	1340	1670	1350	1410	1330
Libya	2210	2120	1080	1050	100
Algeria	1130	1230	100	990	980
Qatar	570	510	340	380	330
Ecuador	210	210	280	270	170
Gabon	150	200	150	160	150
Total OPEC	31350	31690	17330	19740	19360

Table 4.2(*b*): Top twenty non-OPEC producers.

Thousand barrels daily	1973	1979	1985	1986	1987
USSR	8760	11750	11980	12390	12570
United States	10950	10140	10580	10230	9910
Mexico	530	1620	3010	2750	2890
China	1000	2120	2520	2630	2680
United Kingdom	10	1600	2720	2720	2660
Canada	2120	1770	1720	1610	1690
Norway	30	430	850	900	1040
Egypt	170	530	910	820	940
Australia	420	500	660	590	630
India	150	270	600	640	630
Brazil	170	170	570	610	610
Oman	290	290	500	560	580
Malaysia	90	270	440	500	490
Argentina	430	480	490	470	460
Colombia	160	140	180	300	380
Angola	160	140	250	280	360
Romania	310	250	240	250	250
Syria	110	170	180	200	240
Cameroon	0	30	190	180	170
Peru	70	210	190	180	170
Others	1360	1480	1770	1820	1890
Non-OPEC	27290	34360	40550	40630	41240

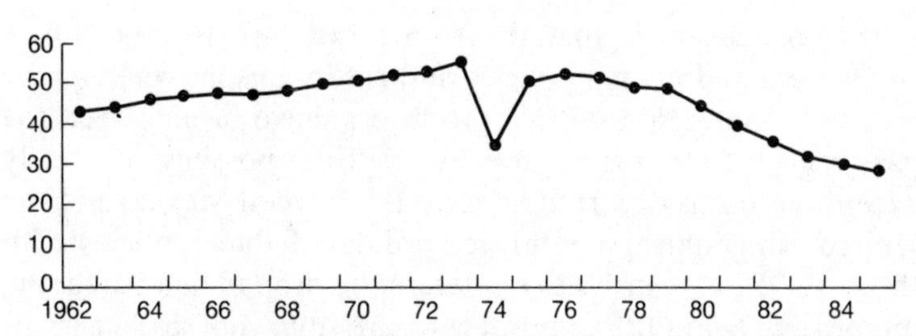

Fig. 4.4: OPEC production as percentage of world total.
Source: *OPEC Annual Statistical Bulletin* (various issues).

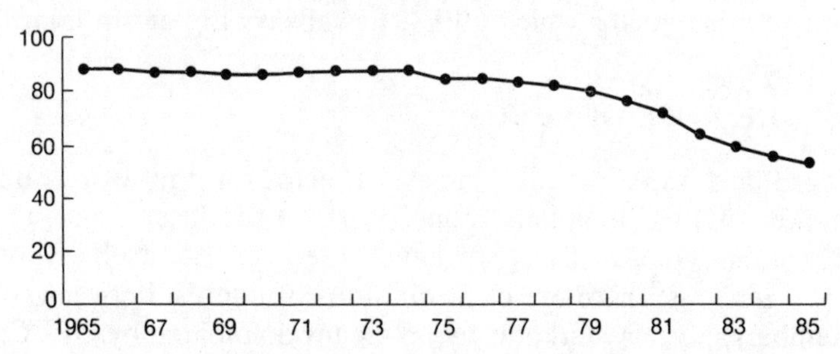

Fig. 4.5: OPEC exports as percentage of world total.
Source: *OPEC Annual Statistical Bulletin* (various issues).

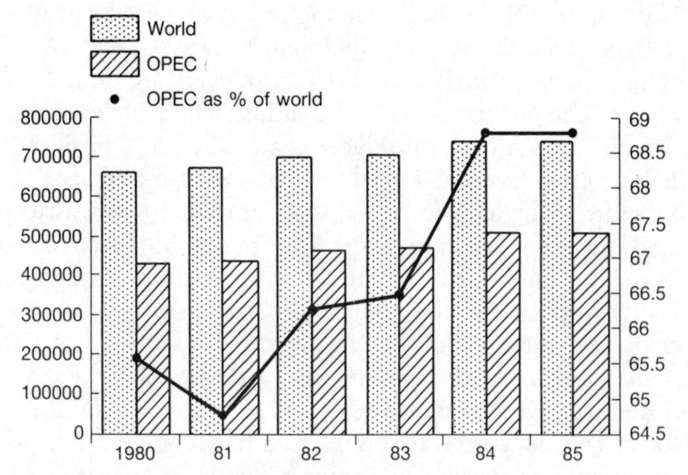

Fig. 4.6: World and OPEC proven crude oil reserves.
Source: *OPEC Annual Statistical Bulletin* (various issues).

OPEC producers is that the former attempt to co-ordinate production and pricing policies and act in unison, whereas the latter do not. Non-OPEC producers have usually reacted passively to the prices set by OPEC: possibly the only exception to this occurred in early 1983, when Mexico and the United Kingdom apparently co-ordinated their policies with those of OPEC in order to moderate the drop in nominal crude oil prices. Non-OPEC producers are thus not a source of change in the world oil market, but their ability to expand output in response to higher prices, and so cut away OPEC's demand, is a factor which OPEC has always to bear in mind.

4.4 THE ROLE OF OPEC

There are three major classes of actor in the world oil market—OPEC members, non-OPEC oil producers, and oil-consuming countries. As we have noted, OPEC itself is not homogeneous: there are clear divisions of interest between its members. So this market is by no means dominated by OPEC: what happens in the world oil market is best understood as the outcome of several different games that are being played simultaneously. There is a game for influence over pricing policy within OPEC between high-reserve and low-reserve members. Then OPEC as a whole must always glance over its shoulder to watch what non-OPEC producers are doing.

Finally, the governments of the industrial countries, several of which are major oil-producers, are also in a position to influence price levels and price movements. No one agent can hold another single agent responsible for what happens to oil prices: all are, whether they like it or not, collectively responsible, and have contributed to the evolution of oil prices.

In the next chapter we analyse this movement of oil prices over the last quarter century, and try to indicate the roles of the various participants in the market. In particular we try to indicate clearly the impact that OPEC has had in the oil market. The evidence is that it is a significant but not dominant participant, very much in keeping with the theoretical arguments of this chapter.

NOTES

1. A conventional measure of monopoly power, known as the 'Lerner Index', is the difference between price and marginal cost expressed as a fraction of price. The model underlying this is of a market for a reproducible commodity produced under conditions of non-increasing returns: in a competitive market, such a commodity will be priced at marginal cost. In this case, the difference between price and marginal cost can be used as an index of monopoly power. However, this index is not applicable in resource markets, where even under perfect competition price will exceed marginal cost. See R. S. Pindyck, 'On monopoly power in extractive resource markets', *Journal of Environmental Economic and Management*, 14 (1987), 128–42, and also 'Pouvoir de monopole sur les marches des ressources non renouvables', in Gaudet and Laserre (eds.), 'Ressources Naturelles et Theorie Economique', *Les Presses de L'Université de Laval* (1986).
2. For a detailed review of OPEC's recent behaviour, see Dermot Gately, 'Do Oil Markets Work? Are OPEC dead?', *Annual Review of Energy*, vol. 14.
3. A more detailed development of a general equilibrium analysis of this area can be found in Chichilnisky, 'Oil prices, industrial prices and outputs: a simple general equilibrium analysis', discussion paper, Columbia University Department of Economics (1980) and Chichilnisky and Heal, *The Evolving International Economy* (Cambridge University Press, 1987).

   Other relevant references are Chichilnisky, 'A general equilibrium theory of North-South trade', ch. 1, vol. II in Heller, Starr and Starret (eds.), *Essays in Honour of Kenneth J. Arrow* (Cambridge University Press, 1988): Chichilnisky, 'International trade in resources: a general equilibrium analysis', in R. McKelvey (ed.), *Environmental and Natural Resource Mathematics* (Proceedings of Symposia on Applied Mathematics, American Mathematical Society, Providence, Rhode Island, 1985), pp. 75–125: and Chichilnisky, 'Prix du petrole, prix industriels et production: une analyse macroeconomique d'equilibre general', in G. Gaudet and P. Laserre (eds.), *Ressources Naturelles et Theorie Economique* (Quebec, Les Presses de l'Universite de Laval, 1986), pp. 26–56.

5

Oil Price Movements: Interpretations and Predictions

5.1 BACKGROUND

Fig. 4.1(*a*),(*b*) in Chapter 4 shows movements in the price of crude oil, measured in constant US dollars. Clearly this series changed its behaviour in the early 1970s. Prior to then, the constant-dollar (and sometimes even the current-dollar) price of oil was drifting slowly and steadily downwards. Over this period, the oil market was seen by most commentators as being in a state of perpetual surplus, a state which was expected to continue. Then, in the early 1970s, matters clearly changed. Note that although the big price rise that made OPEC famous happened in 1973, the change in the trend of prices in fact occurred before then: the price of crude oil in fact started rising in 1970, and by 1973 was increasing at historically unprecedented rates. From 1973 to the present, constant-dollar oil prices have moved erratically: they have been typified by occasional sharp increases interspersed amongst periods of constancy or downwards drift. The drop in prices after 1983 introduced a new element.

In this chapter, we use the analysis of the previous four chapters to provide an explanation of this behaviour of prices, and to provide tentative forecasts of future development.

5.2 BEFORE 1970

Up to 1970, the price of oil fell: why? Several factors can contribute to an explanation—discoveries of large new deposits, negative real interest rates, expectations about price movements, and the role of the major oil companies, who were the main purchasers.

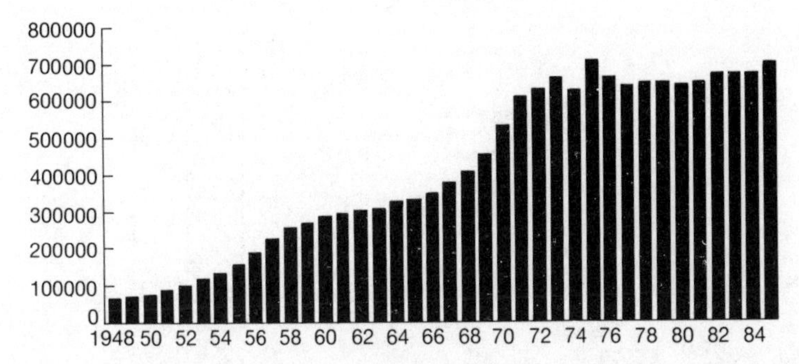

Fig. 5.1: Estimated world proven reserves of crude oil: millions of barrels.

Source: *OPEC Annual Statistical Bulletin* (various issues).

Fig. 5.1 shows the evolution of proven reserves of crude oil from 1948 to 1985. By 1970 these stood at nearly twice their level in 1960, in spite of the high levels of oil consumption in the 1950s and 1960s. This reflects a series of major discoveries: Table 5.1 shows recent trends in discoveries. Over the period to 1970, the balance between supply and demand was shifting, with supply rising faster than demand. This inevitably led to a downward pressure on prices.

A variable which our theoretical analyses suggested should be important in determining oil price movements is the real rate of interest. Chapter 1 reviewed both theoretical arguments and empirical evidence on this relationship. Fig. 5.2 shows real

Table 5.1: Trend of world oil discovery and production (million metric tons).

Time period	Average annual production	Average annual discovery*	Discovery/ production ratio
1950–9	000.1	3821.0	0.04
1959–69	1475.2	4363.8	2.96
1969–79	2751.8	4015.1	1.46
1976–80	3070.4	2792.8	0.91

* Yearly average of reserve addition plus production during the period.
Source: *World Energy Outlook*, 1982.

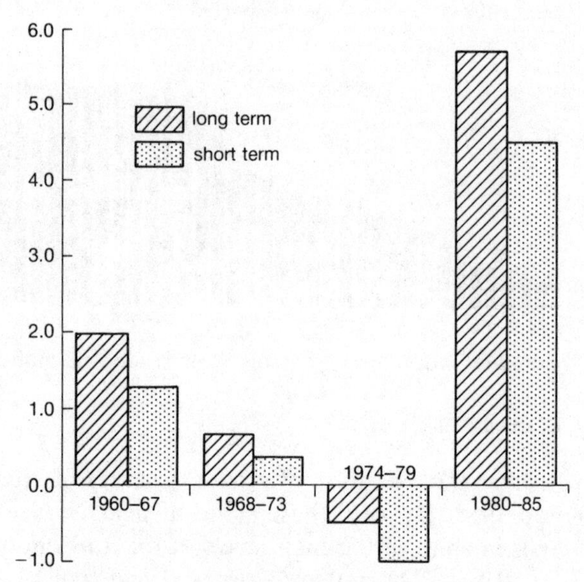

Fig. 5.2: Real interest rates 1960–7, 1968–73, 1974–9, 1980–5, per cent per annum.

Source: OECD historical statistics, 1987.

long-term and short-term interest rates in the USA over the period 1960 to 1985. It is difficult to argue a tight correlation with oil price movements: this would anyway have to be the subject of a complex econometric study. However, the period when oil prices rose sharply (1972–8) saw a sharp drop in real interest rates, and the period when prices fell (1983–8) was characterized by a rise in real interest rates to a record level. There is therefore a case for the relationships which the theory predicts.

Another important factor is that price movements in many commodity markets can be self-reinforcing, at least over limited periods of time. The reason is that in these markets, a very important role is played by the expectations held by traders about the prices that will rule in future periods. Producers have to decide whether to produce and sell now, or whether to delay until a future date. Obviously, they will be tempted to delay if in the interim they anticipate a significant rise in prices, so that by postponing their sale they can realize a capital gain. Conversely, they will sell now, rather than later, if they expect a significant drop in prices, for by selling now they avoid a capital loss. Now, traders' expectations of how prices will move in the near future are strongly influenced by their experience in the recent past: there is a tendency to extrapolate recent trends into the future, at least unless there is some obvious reason for circumstances to change. So a history of prices increases will induce expectations of more of the same, and will encourage sellers to hold their product off the market in anticipation of capital gains. Conversely, a history of falling prices will tend to boost supply as sellers rush to avoid capital losses.

On the consumer side, exactly the opposite arguments apply. If prices are expected to rise, for example because they have been rising, then consumers clearly will buy sooner rather than later, raising demand. An expectation of price falls will have the opposite effect.

Putting together the consumer and producer sides of this story, it is clear that a period of falling prices can be self-reinforcing. It encourages producers to increase supply to avoid capital losses, and it encourages consumers to reduce demand and buy later: both of these tendencies put further downward pressure on prices. A symmetric argument applies to price increases: producers will hold off the market, and consumers will anticipate future needs, reinforcing the upward pressures.

Finally, there is an important institutional factor: the role of the international oil companies. Most non-economists are familiar with the economist's concept of monopoly—a single or dominant seller of a good. Less well-known is the mirror-image concept of monopsony—a single or dominant buyer of a good. Monopsony power is the power of a single buyer to hold the price of a good down: he or she faces no competition for the good, the sellers have nowhere else to turn, and the price may be forced very low.

Prior to 1970 the major Western oil companies acted as a buying cartel: they co-ordinated their crude oil buying strategy just as OPEC now co-ordinates its selling strategy. Because the oil market was anyway a buyer's market, and because the oil-producing countries were still largely politically subservient to the West, the oil companies were able to use their market power to control the price of crude oil, even to the extent of on several occasions forcing cuts in the nominal price on the selling countries.

It seems, then, that it is not difficult to explain why prices were falling prior to 1970: supply was rising faster than demand, real interest rates were low or negative, expectations were of a continuing surplus and of downward pressure on prices, and the major oil companies formed an active buying cartel which was unmatched by an equivalently active selling cartel. In short, the world was a very different place then.

5.3 THE EARLY 1970s

A well-established trend changed in the early 1970s. As one might expect, a number of factors contributed to this. They included a sharp change in traders' perceptions of the supply-demand balance, leading to changes in expectations, and a change in the relative monopsony and monopoly powers of buyers and sellers respectively. There was also a transfer of control and ownership over major Middle Eastern oil reserves.[1] Once a new trend was established, the self-reinforcing tendencies already mentioned were there to continue it.

Recall from Fig. 5.1 and Table 5.1 that in the 1960s and 1970s there was a decline in the rate of discovery of oil reserves. Fig. 5.3 shows that over this period energy demand was rising (Fig. 5.4 shows a breakdown of OECD energy demand). By 1982 proven reserves of oil were only 25 per cent greater than in 1970 (see Fig. 5.1), in spite of the great increase in prices between 1970 and 1982 and the very extensive exploration and development activities carried out in the interim. There were few really major discoveries during this period, and much of the addition to reserves consisted of bringing into production known but previously uneconomic deposits of oil. Oil was no

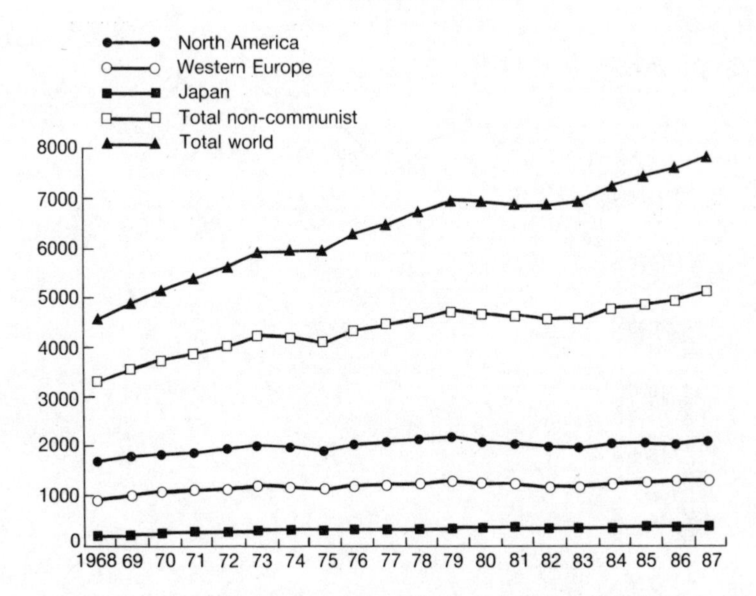

Fig. 5.3: Energy consumption by region in millions of tons of oil equivalent.

Source: British Petroleum, *Annual Review of World Energy*, 1988.

longer becoming more abundant, and began to get scarcer. Perhaps just as important, the world became very conscious of this fact. A series of widely publicized studies drew the public's attention to the constraints that resource-scarcity could impose on economic growth. The earliest of these were the *Limits to Growth* and the *World Dynamics* studies.[2] In retrospect they appear exaggerated and inaccurate; at the time, however, they had a substantial impact, as they emphasized for the first time in the post-war era the fact that mineral resources could in principle be depleted. Both buyers and sellers became much more aware of the scarcity and strategic value that oil would come to have.

In addition to the changing balance between discoveries and consumption, and the growing awareness of possible shortages, another factor contributed to changing the tone of the market: this was the rapid passage of the US economy from oil self-sufficiency to substantial dependence on imports, a product of

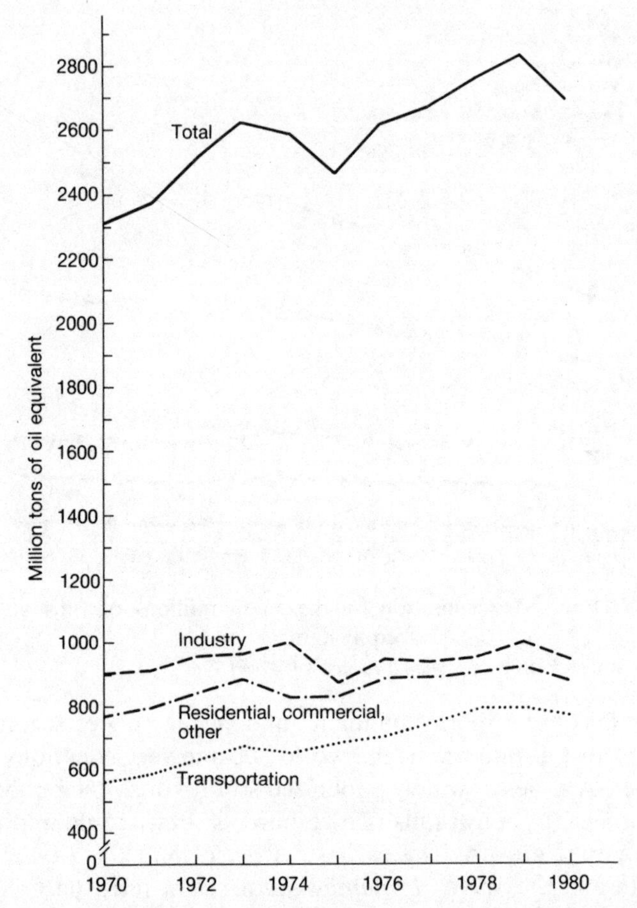

Fig. 5.4: Sectoral composition of final energy demand in OECD countries, 1970–80.

Source: IEA, *Energy Balances of the OECD Countries* (various issues).

rising US oil consumption and falling US oil production. Fig. 5.5 shows the growth of US oil imports in the late 1960s and early 1970s. US imports grew from about 1 million barrels per day (mbd) in 1971 to 6 mbd in 1977. The US remained one of the world's principal producers (see Fig 4.2 (*b*)), but experienced a rapid growth in demand. A substantial new customer entered the world oil market.

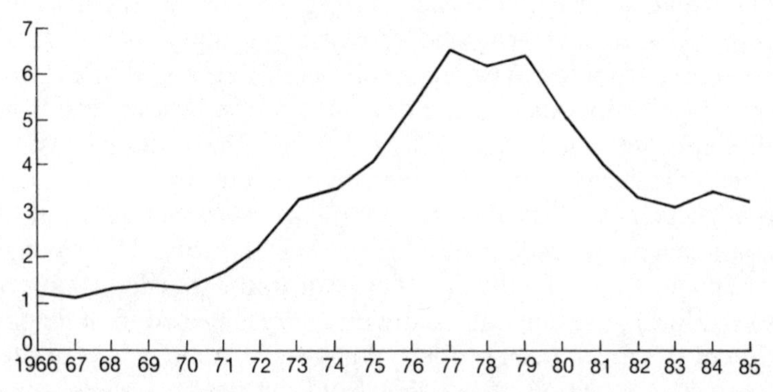

Fig. 5.5: US crude oil imports (millions barrels per day).

The balance between oil supply and oil demand in the international market changed in the early 1970s. This alone would have sufficed to change the trend of prices. However, during this period there were also important changes in Middle Eastern politics, and they acted to reinforce this newly emerging trend. The region began to develop an increasing degree of political independence of the West, and to assert a growing measure of control over its own economic resources. Nasser's nationalization of the Suez Canal in 1956 was a foretaste of this: Gadaffi's rise to power in Libya marked the first time that the trend really affected the oil market. The oil countries experienced a growing desire to control their own economic destinies, just as conditions in the oil market changed in such a way as to allow them to do so. Perhaps this was no coincidence: the desire may have been there all along, but suppressed as long as it looked unattainable.

This desire for greater economic independence manifested itself through measures designed to give host governments more control over production from wells on their territory, and a much larger share of the revenues from oil sales. In the late 1950s and 1960s, production and pricing decisions were controlled almost entirely by the major oil companies, with a very minor role for the governments of the host countries. Attempts to assert local control over oil fields, such as that of the Iranian prime minister Mossadeq to nationalize oil companies in Iran in 1948, provoked prompt governmental responses from the Western countries. The transfer of control over production and pricing decisions to the host countries in the late 1960s and early 1970s led to a cutback in production rates. The new controllers were more concerned than their predecessors to conserve the economy's resource base, and to assure a long-run supply of oil from their territory. This change was compounded by the fact that prior to the transfer of power to the local governments, many oil companies had in anticipation of this event been producing at very high rates, in order to remove as much oil as possible from the deposits whilst they were still under their control. There was thus a sharp reduction in the production of some OPEC members, reinforcing the other upward pressures on oil prices. The desire of the oil countries for greater economic independence was also expressed through a more aggressive role for OPEC: the major oil exporters realized that a common front and a co-ordinated pricing policy could greatly increase their bargaining power.

We now have a picture of a tightening oil market, switching from a buyer's to a seller's market, and sellers eager to get a better deal than they had before. The price trend switched, and the new trend reinforced itself by the processes we discussed before. In the early 1970s oil ministers were often heard declaring that oil in the ground gave a better return than money in Western banks, meaning that the rate of price appreciation of oil gave them a strong incentive to delay selling and enjoy capital gains. And the sharp drop in real interest rates in the early 1970s may have given this process further impetus.

5.4 THE PERIOD 1973–83

In the short run, the demand for oil is insensitive to price. So the increase of 1973 had little immediate effect on demand. However, when it became clear that the new price levels would be sustained, and that oil-pricing policies had entered a new era, the higher prices set in motion a number of processes that have still not been completed. These are the kinds of processes described in Chapter 3: development of more fuel-efficient vehicles and aircraft, or more fuel-efficient industrial processes, and of building and heating technologies oriented towards minimizing heating and cooling costs. At the same time, the increased relative prices of energy-intensive activities began to shift consumer demand away from these and towards alternatives whose costs were less closely tied to the price of energy. None of these factors affected the demand for oil much before the late 1970s: then their consequences began to be felt, leading to progressive reductions in the energy-intensity of the goods and services produced and consumed in the industrial countries. A much-quoted statistic on this subject is that the amount of energy used in producing $1,000-worth of constant-dollar GNP in the USA fell by 38.9 per cent from 1973 to 1983. This was a result of switching to more energy-efficient technologies, and mostly of demand patterns changing away from energy-intensive products and services. Most of this striking drop in energy use occurred in the period 1979–83.

The 1973 price increase also set in motion important long-run forces on the supply side. In particular, it initiated a major rise in non-OPEC oil production, as noted in Chapter 4. The mechanism by which this occurred was very straightforward: at higher prices, the rewards of oil exploration and development were greatly increased, leading to a dramatic rise in exploration activities. At the same time, known oil deposits whose extraction costs were too high for them to yield a profit at $1.50 per barrel, became quite attractive propositions at $12 per barrel. So new deposits were discovered and brought into production (as for example in Mexico and the North Sea): and known but previously submarginal deposits were brought into production. All of this takes time: it may be five years from the discovery of a new oil field to the time when it starts producing, and indeed if the terrain is difficult, as in the North Sea or in Alaska, the delays can be longer.

So the 1973 price increase had little immediate effect on demand or supply, but set in motion forces which a decade or so later would lead to a squeeze in the market: it set in motion a gradual build-up of non-OPEC supply, and a gradual decline in demand.

It is these events that were primarily responsible for the downward pressure on prices in the early 1980s. This was not, as is sometimes argued, primarily caused by the recession in the industrial countries. This is easily appreciated by some simple calculations. Take the income elasticity of demand for oil to be 0.8, as suggested in Chapter 3. At the worst point of the OECD recession, levels of national income declined less than 5 per cent from their previous peaks: if the income elasticity of demand is 0.8, this would lead to a drop in demand for oil of at most 4 per cent.

Estimating the impact on demand of higher crude oil prices requires a little more patience. In 1973 crude oil prices rose 400 per cent, from $3 per barrel to $12 per barrel. This contributed between 20 cents and 25 cents extra to the cost of a gallon of petroleum products, leading to an increase in their consumer prices of between 40 and 50 per cent. For simplicity we will work with the 50 per cent figure. Table 5.2 shows the drops in demand that would result from a 50 per cent price rise, for various values of the price elasticity of demand. A short-run price elasticity of about −0.2 would imply that the impact effect of the 1973 price rise was a demand drop of about 10 per cent. If the long-run price elasticity is of the order of −0.5, then in the long run there would be a demand reduction of about 25

Table 5.2:

P-elasticity	Drop in demand %
−0.1	5
−0.2	10
−0.3	15
−0.4	20
−0.5	25
−0.6	30
−0.7	35
−0.8	40
−1.0	50

per cent. If the long-run elasticity were of the order of −1.0, then we would expect the 1973 price to lead to a drop in demand of about 50 per cent. In fact 1983 might be too early for the full effects of the 1973 price rise to be felt. If two-thirds of this effect had been felt by 1983, this would account for a demand drop of either 17 or 35 per cent, depending on the elasticity figure chosen: this is clearly much greater than the roughly 4 per cent attributable to the OECD recession. In fact to this 17 or 35 per cent owing to the 1973 price rise, we must add the short-run consequences of the 1978–9 price rise. Then higher prices can easily account for a demand drop of over 20 per cent from its peak, and quite possibly nearer 40 per cent.

The decade 1973–83 was one in which the world began to adjust to higher oil prices. The adjustment was not completed: some of it will still be under way at the end of the century. Demand dropped of the order of 20 per cent or more because of higher prices, and of the order of 5 per cent because of the recession in industrial countries. Expansion of non-OPEC production capacity contributed an extra 18 per cent to the world's crude oil capacity. Not surprisingly, there was a strong downward pressure on prices, and major production cutbacks by OPEC were needed to stem the decline in the current US$ price of crude oil. It should be noted that this $4 drop in the US price in 1983 did not lead to equal drops in oil prices in other currencies. It occurred during a period of rising exchange rates between the dollar and the currencies of other industrial countries, so that in European currencies or in Japanese yen the price of crude oil stayed constant or even rose slightly in 1983. Nevertheless, the drop in the US$ price would have raised demand in the US albeit by rather little.

One final remark is in order on the 1973–83 decade in the oil market. Towards the end of this decade, as we have seen, the demand for OPEC oil fell substantially. Total demand declined, and non-OPEC production rose. There are two ways in which a producer can respond to a decline in demand relative to supply—by cutting back output, or by reducing price. As the demand for oil is very price inelastic, at least in the short run, OPEC unquestionably stood to lose less in terms of current revenue by reducing output to match the lower demand, than by reducing price enough to raise demand back to their full-capacity output level.

To understand this, consider the following numbers. Suppose that at present prices, the demand for OPEC oil has fallen 20 per cent, and that the short-run price elasticity of demand for oil is −0.3. Then if OPEC simply cuts back output by 20 per cent, keeping the price constant, its revenue falls by 20 per cent. Suppose on the other hand that it tried to boost demand 20 per cent back to the original level by cutting price. A price reduction in excess of 60 per cent would be needed, leading of course to a revenue drop of the same size. So by cutting output much more than price, OPEC clearly made their best response to the changed market conditions. A major distinction between the behaviour of OPEC members, and of non-OPEC producers, was that the former reduced output while the latter worked to expand output as fast as possible.

In the previous chapter we analysed OPEC's role in the world oil market, and the extent to which it can be said to exercise market power. The ability to ensure that a demand reduction was met by a cut in production rather than in prices, would probably be the most important indication of OPEC's power, and of what differentiates the oil market from other commodity markets. Without a cartel framework to negotiate reduced production quotas and to provide some assurance that they would be observed, the excess supply would almost certainly have led to price rather than output reductions. In fact this is just what happened with other commodities. In the early 1970s, their prices moved very much in parallel with oil, rising sharply. However, as demand fell because of recession and substitution, their production levels remained constant and their prices fell sharply, reaching historical lows in real terms by the early 1980s. Fig. 5.6 shows the real price movements for several other primary commodities. That oil avoided their fate in the 1980s, although it had shared it during similar periods in the 1950s and 1960s, is a clear measure of OPEC's impact on the oil market.

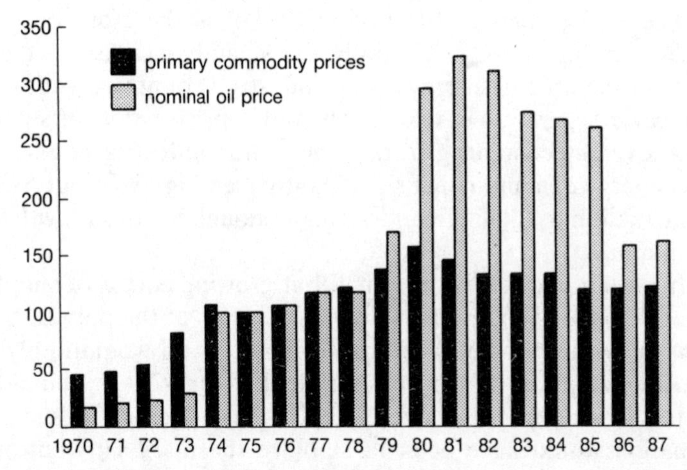

Fig. 5.6: Indices of non-oil primary commodity prices and of crude oil prices.

Source: *UN Monthly Bulletin of Statistics.*

5.5 THE PERIOD 1983–8

The period from 1983 to late 1989 saw a fall in the real price of oil almost as spectacular as the rise in the 1970s (see Fig. 4.1(*a*),(*b*). In many respects, exactly the same factors were responsible for both events, though of course acting in reverse in the period 1983 to 1989.

During this period, non-OPEC supply of crude oil continued to grow (see Table 4.2(*b*). The growth was steady and not spectacular, and was set in motion by the price increase of the 1970s, which initiated major rounds of exploration, and of investment in enhanced recovery. Many projects in these areas have lead times of the order of a decade, so that by the mid-1980s the consequences of the 1970s price increases for supply were steadily making their way onto the market.

On the demand side, the scene was also set by the 1970s price increases, which, as we have already noted, set in motion a long-term change in oil use patterns and demand characteristics which even at the time of writing has not been fully implemented.

The background to the period 1983–9 is therefore one of steadily rising non-OPEC supply and steadily increasing efficiency in the area of energy use in industrial countries, with the downwards pressure of this on demand approximately offset by their overall economic growth. Only in non-industrial countries was energy demand rising significantly (see Fig. 5.7), but their contribution to total demand was not enough for this growth to be important.

In such a market, with a small but growing excess of supply over demand at the prices ruling at the start of the period, the low short-run price elasticity of demand for oil would imply a substantial price fall to equilibriate the market. As already noted in section 5.4 above, OPEC's best strategy in such a situation would have been to stabilize the market by cutting output rather than allowing a decline in price to make the adjustment. As we remarked in section 5.5, with a very low short-run price elasticity of demand, the revenue loss is much less if output rather than price bears the burden of adjustment.

OPEC of course tried once again to follow this strategy,

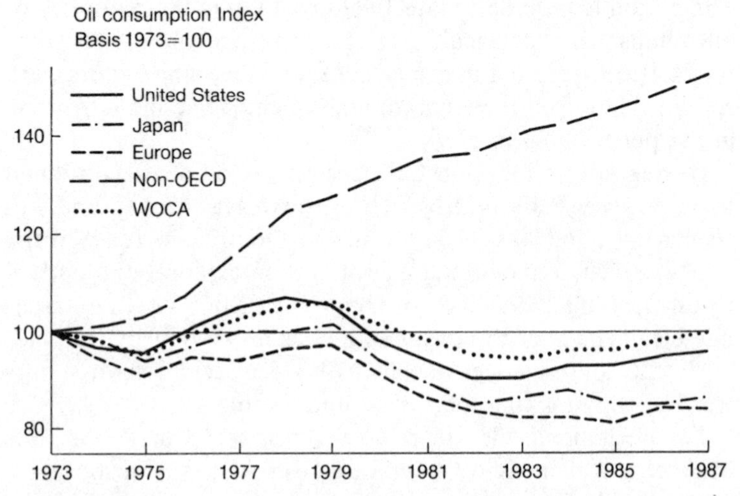

Fig. 5.7: Indices of oil consumption by region. WOCA = world outside communist area.

Source: Shell Briefing Service, *Energy in Profile*, number 3 (1988).

which it had followed successfully in the early 1980s, but failed. They were not able to ensure the implementation of the production quotas which they agreed. A number of OPEC member countries acquired refining and retail distribution capacities in the industrial countries, which they used to ensure outlets for their above-target output. In addition, Saudi Arabia was unwilling to continue playing the role of a 'swing producer' who adjusted their output to compensate for over-production by others. In fact, by the autumn of 1988 Saudi Arabia was exceeding its official OPEC target by about 1.5 million barrels daily, or over 25 per cent. There was a presumption in the industry that this was part of a process of seeking an advantageous position from which to negotiate a new set of OPEC quotas.

To put this remark in perspective, it should be noted that by the mid-1980s, OPEC's share of world oil production had fallen to about one-third (see Fig. 4), having been as high as 55 per cent in the early 1970s. Likewise, OPEC's share of world oil exports had fallen from 90 per cent in 1973 to 60 per cent in 1987 (Fig. 4.5). So OPEC's failure to control the market in the mid-1980s should not be attributed only to internal dissension: the fifteen years from 1973 on saw a major, albeit gradual, decline in OPEC's role and importance in the world oil market. This was an inevitable consequence of the forces set in motion by the price increases of the 1970s.

At the time of writing, it is too early to tell whether OPEC will succeed in establishing and maintaining a price of about $18, or whether prices in the low teens will rule in 1990. Prices of £14–15, if sustained for several years, could boost demand by 5–10 per cent, offsetting some of the impact still feeding through from the 1970s price increases, and providing preconditions for another round of sharp price increases.

5.6 THE 1990s

How is the oil market likely to develop? In the medium term—say the next decade—it seems safe to predict a continuation of the trends just described. Non-OPEC production will remain high. Demand will continue to decline as a long-run consequence of the 1973 price increase, with a drop of at least 10 per cent over the next five years. The 1978–9 increase will also have an effect: this raised retail prices by about 30 per cent, and so might contribute at least a further 15 per cent in round numbers to the drop in demand, taking effect over the next ten years. The only offsets to these downward pressures will come from expansion in the oil-consuming countries, and from the effect of the price reduction in the 1980s on demand. Economic expansion in the industrial countries at 2 per cent per annum, maintained for ten years, could add a little less than 20 per cent to annual demand by the tenth year.

The picture that emerges over the next decade is thus of considerable uncertainty: price increases will continue to feed through to demand, leading to a drop of about 25 per cent: industrial expansion of 2 per cent per annum and the long-run effects of the 1983–8 price fall may offset this. There may be a slight reduction in non-OPEC production—for example as the North Sea fields pass their production peaks. Rapid industrial expansion in some of the middle-income developing countries may strengthen the trend of demand. Alternatively, more rapid expansion of the industrial countries could lead to a net increase in demand. However, none of these will produce real upward pressure on prices, because of the large margin of unused capacity in OPEC.

The expected outcome is thus essentially a continuation of the present. However, a forecast of this type can easily be upset. Continuation of the present requires that OPEC members keep their production levels substantially below their theoretical maximum—Table 5.3 shows the 1980 outputs of OPEC members and their theoretical maximum output levels. For countries which are very short of foreign exchange, the temptation to raise production will be very great. If one country raised output, and this move were not followed by others, then the market price would move little and the expanding country would earn more foreign exchange. In fact such a move would almost certainly be followed by other

Table 5.3: OPEC crude oil productive capacity (Mbd).

	1981 actual crude oil output	1986* actual crude oil output	Maximum sustainable capacity 1985
Algeria	0.8	0.7	0.9
Ecuador	0.2	0.3	0.1
Gabon	0.2	0.3	0.1
Indonesia	1.6	1.3	1.3
Iran	1.3	2.0	4.0
Iraq	0.9	1.9	3.5
Kuwait	1.1	1.2	2.5
Libya	1.1	1.3	2.0
Nigeria	1.4	1.5	2.2
Qatar	0.4	0.3	0.5
Saudi Arabia	9.8	4.8	9.5–10.5
United Arab Emirates	1.5	1.3	2.0–2.4
Venezuela	2.1	1.6	2.4
TOTAL OPEC	22.5	18.3	31.6–33.0

* *OPEC Annual Statistical Bulletin* (1986).
Source: *World Energy Outlook*.

countries, and the resulting extra output would initiate another round of price reductions. In the end, all oil-producers would be worse off in terms of foreign exchange. At the time of writing, it is too early to tell whether OPEC will succeed in holding the line on production ceilings for a substantial number of years. Obviously, they would be aided in this by a relatively rapid economic expansion of the industrial countries, which would take some of the downward pressure off prices.

If one looks further ahead than the next decade, a rather different picture emerges. By the mid-1990s, we will have seen most of the consequences of the price increases of the 1970s. Demand patterns will have stabilized with the continuing downward pressure removed. On the supply side, unless there are significant new discoveries, non-OPEC production will be past its peak. Production from the North Sea and Alaska will be declining, and it is possible that the same will be true of Mexico and the USSR. Some OPEC members will also have declining output levels—e.g. Iran and Algeria. So demand will be stabilizing just as production again begins to concentrate in OPEC, and indeed in those OPEC members in the Persian Gulf. In principle this could set the scene for a period of rising prices, though such a forecast is very tentative: a lot of factors could intervene.

One would be a major breakthrough in the field of 'backstop technologies' as discussed in Chapter 2. For example, the development of a process for producing crude oil from unconventional oil deposits (shale, tar sands) conveniently and in the range of $30–$40 (1983 prices), would certainly put a long-run ceiling on prices. One might also speculate about the possibility of finding really important new oil fields, for example in China or elsewhere in the Far East. To have a major effect on prices they would have to be very large—comparable to some Middle Eastern fields—and this seems most unlikely, but cannot be ruled out.

There could also be unexpected developments that exert an upward pressure on prices. One such eventuality would be the complete abandonment of nuclear power programmes in industrial countries. Another would be a substantial reduction in the mining and combustion of coal. Large-scale coal mining has harmful environmental effects and leads to major ecological disruption: the combustion of coal produces atmospheric pollution and acid rain. It is conceivable that environmental concerns could restrict the rapidly growing use of coal, shifting back to oil (or gas) some of the demand that has transferred to coal since 1973. Finally, a rather different possibility is a major political disruption in the Middle East.

Making predictions in the oil market remains, as ever, a risky business. What we have set out above is a benchmark case, showing how matters are likely to develop if there are no major surprises. A few possible surprises have been mentioned, but there must surely be many that have not!

NOTES

1. See for details Chapter 1 (esp. Fig. 1.3) in J. M. Griffin and D. J. Teece, *OPEC Behaviour and World Oil Prices* (1982).
2. The references are to J. W. Forrester, *World Dynamics* (1971) and D. H. Meadows *et al.*, *The Limits to Growth* (New York and London, 1972). A later and more sanguine view of the same issues is in G. Chichilnisky, A. Herrera, H. Scolnick *et al.*, *Catastrophe or New Society* 1976, International Development Research Centre (Ottawa).

6

Oil, Employment and Inflation

6.1 INTRODUCTION

The issue in this chapter is the effect of oil prices on the level of employment, on the rate of growth of productivity, and on the rate of price inflation in the industrial countries. The industrial countries of the OECD experienced a rather abrupt change of macroeconomic regime in the early 1970s: they moved from growth with full employment and price stability to an unappealing combination of slow growth or even stagnation, employment, and inflation. In this 'stagflationary' regime, the stabilization of prices proved possible only at a very high cost in terms of unemployment and foregone output.

It has been widely asserted—more often than not by non-economists—that this switch of regime in the early 1970s was caused by the increase in oil prices. So in discussing the macroeconomic impact of oil prices, we will pay particular attention to claims of this sort. We shall examine the short-run and long-run effects of high oil prices on the macroeconomic situations of the industrial countries, and see in particular whether these might be large enough to be responsible for a major change in economic performance.

6.2 PERSPECTIVE

The value of oil consumption as a fraction of GNP is relatively small in the OECD countries: in the USA it was 1.8 per cent in 1973 and had risen to 5 per cent by 1983. A priori it certainly seems unlikely that a rise in the price of such an input could trigger a major macroeconomic dislocation.

Although inflation rates prior to 1973 were lower than those after, they were nevertheless rising toward 1973: the same is true of levels of employment, as Figs. 6.1 and 6.2 show. So

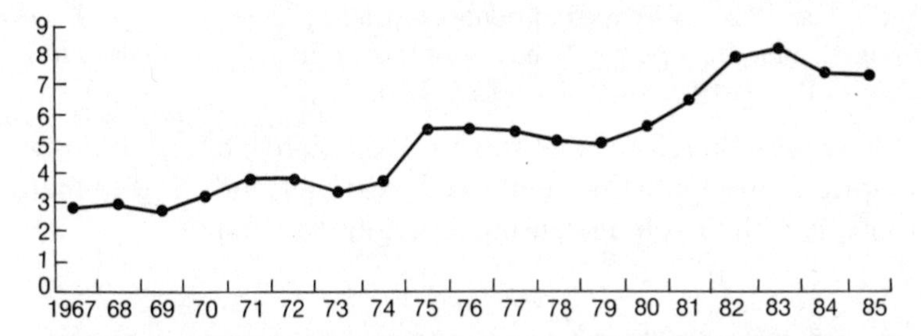

Fig. 6.1: Unemployment level in seven main OECD countries (%).
Source: *OECD Economic Outlook* (December 1987).

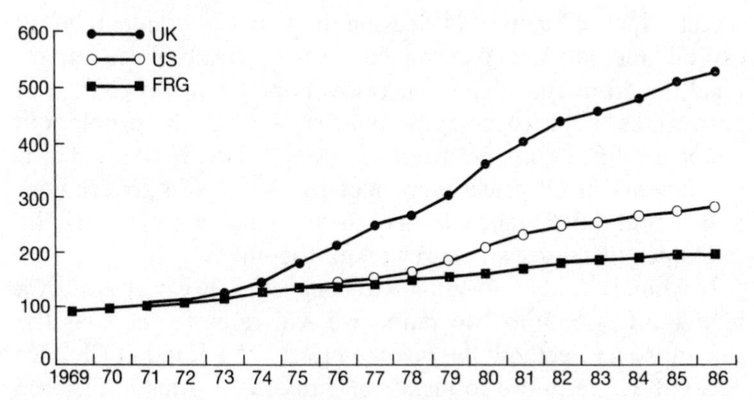

Fig. 6.2: Consumer price index numbers in US, UK, and West Germany.
Source: *UN Monthly Bulletin of Statistics.*

1973 marked not an abrupt change, but a strengthening of existing trends. For example, a 1970 OECD report commented that:

The general price level will have risen this year by at least 5 per cent in most OECD countries: this is more than double the average rate in the early 1960s. And for the first time in a decade, prices in world trade have been rising as fast as the general domestic level.

Looking to the future, its authors added that:

Some signs of a slowdown are now emerging, . . . but there are also disquieting signs that the problem of inflation may have got worse in the sense that, where traditional restrictive policies have been applied, the effect on prices has been less rapid and less long-lasting than in the past.[1]

It seems, therefore, that our problem is to understand how oil prices fitted into and reinforced existing trends, rather than to explain their role in causing a qualitative change.

6.3 OIL PRICES AND THE EMPLOYMENT OF FACTORS

Perhaps the most important of the issues to be studied, is the direct effect of higher oil prices on the levels of employment of capital and labour. We say direct to indicate those effects resulting from the higher oil price alone, with no change in governments' macroeconomic policies. Part of the problem in understanding that occurred in the 1970s is that major movements of oil prices were met by changes in government policy, and the actual changes in the economy are of course the outcomes of these two causes taken together.

It is analytically convenient to break the direct effects of a higher oil price into two parts: we will refer to these as the substitution effect and the income effect.[2] As noted in Chapter 3, a natural response to higher oil prices is to substitute some other inputs for oil in activities in which oil is used, leading of course to an increase in demands for these other inputs. This may take many forms. At its simplest level, it involves walking or bicycling instead of driving a car: this is the substitution of labour for oil. Or it may involve travelling more slowly: taking more time for a given trip is again a way of substituting labour for oil. Maintaining vehicle engines in better tune by more frequent servicing is another method of substituting labour for oil. Of course, capital can also be substituted for oil—by replacing oil-burning equipment with more modern and efficient units, or by insulating buildings, or in many other ways. The redesign of existing procedures to be more sparing of energy will typically involve the use of extra labour and capital—labour to redesign the system, and capital to implement the modifications. Indeed, new capital equipment always has to be manufactured, so that the substitution of capital for oil will usually involve the employment of extra labour as well.

Another component of the substitution process occurs through alterations in the pattern of consumer demand. Higher oil prices lead to a rise in the prices of oil-intensive products relative to others, and faced with different relative prices, consumers will change their buying habits. They will for example take vacations nearer home to avoid travel: more generally they will consume less energy-intensive products (travel, artificial fibers, aluminium) and more of others whose prices are less sensitive to the cost of energy (home entertainment, electronics, food).

In summary, if the price of oil rises relative to the prices of other inputs to production, this will lead to an increase in the demands for those inputs—capital and labour—whose prices have not risen, and which can replace oil in some uses. Through this mechanism, higher oil prices lead to an increase in the demand for other factors of production.

We mentioned above that we can break the impact of higher oil prices into two parts, a substitution effect and an income effect. And having dealt with the former, we can now turn to the latter. This can be described as follows. If people do not change their consumption of oil, a higher price of oil means an increase in expenditure on oil and a decrease in expenditures on other goods and services. Even if oil consumption falls a little in response to the higher price, the same is true. Now, this fall in expenditure on other goods and services will lead to a fall in demand for the inputs used in making them. The higher the price of oil has a deflationary effect, and this naturally lowers the demands for labour and capital.

Matters do not end with the income and substitution effects. There is in addition a recycling effect: higher oil prices imply greater export revenues for oil-producers and so increased imports of goods and services from the oil-consumers, or increased investment in the oil-consuming countries. This mechanism is nicely exemplified by the strong correlation that exists between Middle Eastern oil export revenues, and the value of orders for arms placed by Middle Eastern states with the OECD.[3] A significant fraction of the OECD's oil import bill is returned as revenue from arms sales, from construction sales, or as investment. The magnitudes involved make the effect important: it has been estimated that within three years of the 1973 oil price increase, over 50 per cent of OPEC's extra revenue was being returned to the industrial countries as export demand.[4] Obviously, then, this recycling effect can be important enough to offset, at least in part, the harmful income effect of a higher oil price.

Summarizing, it is clear that an increase in the price of oil leads to two opposing effects on the demand for labour and capital: the substitution effect raises these demands by displacing to other factors some of the demand for oil, and the income effect, net of recycling, tend to lower them. It is possible to set out conditions under which one effect or the other will dominate, though these are, inevitably, quite complex.[5] Roughly speaking, we might expect the income effect to be greater, the greater is the proportion of income currently spent on oil. And this will in turn be greater, the higher is the current price of oil. So an oil price rise might have a stimulating effect on demands for labour and capital if the price of oil is initially low, and the economy uses little oil. It is likely to have a depressing effect at high oil prices and in economies which use oil extensively.

It may be worth making some very tentative comments about the orders of magnitude of the effects discussed here. The income effect of the 1973 oil price rise has been estimated, for the OECD countries, as slightly less than 2 per cent of GNP.[6] Allowing for a little under half of this to be offset by recycling, we see that the net income effect is of the order of 1 per cent of GNP—significant, but not overwhelming. The employment of labour usually falls more than proportionally to GNP so that this could account for a drop of perhaps up to 2 per cent in employment. Sadly there are no estimates, even very rough, of the substitution effect on the demand for labour, so that there is no way of knowing whether the net effect was positive or negative.

As a final comment in this section, note that it has been

concerned with the direct effects of higher oil prices, i.e. those effects which would occur even without any government response. In fact major oil price movements have always led to changes in macroeconomic policies, though the nature of the change has of course varied with the initial macroeconomic conditions and with the government in power. The 1973 price increase was met by accommodating or expansionary policies, which helped to offset the potentially harmful income effects. The 1978–9 increase, on the other hand, was met by a tightening of fiscal and monetary policies, which reinforced the harmful income effects of the higher oil price. These income effects were anyway probably bigger than those of 1973, as the price of oil was much higher in 1979, and we have noted that income effects are likely to be greater at higher initial prices. Putting these observations together, it is easy to see why the 1978–9 oil price increase should have been associated with a more severe recession than the 1973 increase. Incidentally, one authoritative study has estimated that the loss of output and employment from the restrictive monetary policies of the USA and the UK in 1979 was at least comparable with the losses from the direct effects of the higher oil price.[7]

6.4 OIL PRICES AND PRODUCTIVITY

'The productivity slowdown' has occasioned an extensive literature. The substance of this is that, in the USA in particular, the rate of growth of output per head has declined over the last two decades. This deceleration of productivity growth is seen as a possible cause of the losses of competitive position in the world markets, and thus of consequent declines in profitability and employment. Could higher oil prices affect productivity or its rate of growth?

'Productivity' is used to denote output per unit of labour employed, i.e. it measures the productivity of labour and not of any other factor. The main point to note then is that if the price of oil increases, and consequently individuals and firms use less oil and more labour to perform a given task, then as output is constant and employment rises, there is a decline in productivity. This decline is a direct and natural consequence of the process of factor substitution discussed in the last section.

A rough estimate of orders of magnitude might again be interesting. The effect we are trying to isolate is just the substitution effect on the demand for labour. We commented in the last section that there are no good estimates of this, but we can still carry out some thought-experiments. It was argued above that the income effect of the 1973 price increase, net of recycling, might have reduced employment by about 2 per cent. Suppose, for example, that the substitution effect had just outweighed this, leaving the level of employment constant. Then the substitution effect would have raised employment 2 per cent at a constant output level, and so lowered productivity by 2 per cent. In other words, we could account by this argument for a drop in productivity of the order of 1 per cent per annum for each of two or three years, but not for much more. This is approximately the observed departure from trend in the 1970s.

Another potentially important point is that in the long term, productivity growth depends upon investment and the installation of new capital equipment. If higher oil prices contributed to a decline in investment in newer and more productive equipment, then this could provide an indirect mechanism through which they could contribute to the productivity slowdown. The evidence on this issue is ambiguous. Increases in the price of oil have been associated with declines in corporate profits and in non-residential investment, but it is difficult to disentangle the influence of oil prices from the influences of other factors.

6.5 OIL PRICES AND INFLATION

At first sight, inflation is the macroeconomic phenomenon that is most obviously and naturally linked with higher oil prices. In fact appearances are deceptive: it is no easier to quantify the inflationary impact of higher oil prices than it is to quantify their impacts on employment or on productivity. In the case of inflation, it is rather important to distinguish between the impact effect and the long-run effect. The impact effect is just the immediate effect on the general price level. But because of the wage–price spiral, the process by which wage claims react to price changes to maintain or raise real wages, an initial price increase can lead to wage increases, cost increases, further price increases, and so on. The long-run inflationary effect of higher oil prices is then the impact on the general price level, taking all of these feedbacks into account.

It is easy to get a rough idea of the size of the inflationary impact effect of the 1973 oil price rise. Prior to 1973, the OECD countries spent of the order of 1 per cent of GNP on oil. In 1973 the price of oil approximately trebled. So the increase in their oil bill was of the order of 2 per cent of GNP—just the amount of OECD GNP transferred to OPEC, and the amount of the income effect discussed in section 6.3. If output levels and input levels remained constant, and cost increases were passed on as price increase, then this rise in the oil bill of 2 per cent of GNP would raise GNP by 2 per cent. Obviously, under the assumed conditions of constant outputs, this would reflect a 2 per cent inflation. So the impact effect of the 1973 price rise on the price level would have been of the general order of a 2 per cent increase—it could have been 1 per cent or 3 per cent or even 4 per cent, but it would have been in this range.

To estimate the long-run effect, it would be necessary to know how much this impact effect influenced wage claims, and influenced general expectations of a continuing inflationary process. Several studies[8] have put the long-run effect at about three times the short-run, i.e. at about 6 per cent when all the feedbacks between wages and prices are taken into account. These six percentage points would have been added to the general price level over three or four years, contributing perhaps 2 per cent per annum to the rate of inflation during those years. As inflation rates were then between 5 and 10 per cent annum, this represents a noticeable increase but not a dramatic one.

6.6 SUMMARY

Higher oil prices have affected employment, productivity and inflation: they seem on balance to have affected them in unwelcome directions. But the effects have certainly not been overwhelming: indeed, the effect on employment could have been very small, though those on productivity and inflation were more noticeable. All of these estimates are of course very tentative, as the calculations leading to them have omitted a lot. In particular, they have omitted the effect of oil prices on profits. As profits may influence investment, which in turn influences output and productivity growth, it is conceivable that this is a serious omission. Corporate profits were certainly low in the years after 1973, and non-residential investment fell sharply. This reduction in investment could have lowered employment and, by reducing productivity growth and so the scope for absorbing wage increase, have contributed to inflation. To what extent higher oil prices did cause lower profits, and thus lower investment, is just not clear. Accurate numbers on these effects would in all probability not change the general assessment that higher oil prices were only one of a number of factors contributing to the emergence of the macroeconomic crisis of the OECD countries in the 1970s, but it would be reassuring to be certain of this.

NOTES

1. Taken from *Inflation—the Present Problem* (OECD Paris, December 1970).
2. The analysis of this section follows closely that of Chichilnisky, 'Oil Prices, Industrial Prices and Outputs: a Simple General Equilibrium Analysis', discussion paper, (Economics Department, Columbia University, 1980).

   Other relevant references are Chichilnisky, 'A general equilibrium theory of North–South trade', ch. 1, vol. II in Heller, Starr and Starret (eds.) *Essays in Honour of Kenneth J. Arrow* (Cambridge University Press, 1988); Chichilnisky, 'International trade in resources: a general equilibrium analysis', in R. McKelvey (ed.), *Environmental and Natural Resource Mathematics* (Proceedings of Symposia on Applied

Mathematics, American Mathematical Society, Providence, Rhode Island, 1985), pp. 75–125; and Chichilnisky, 'Prix du petrole, prix industriels et production: une analyse macroeconomique d'equilibre general', in G. Gaudet and P. Laserre (eds.), *Ressources Naturelles et Theorie Economique* (Quebec, Les Presses de l'Universite de Laval, 1986), pp. 26–56.

3. The connection between oil export revenues and arms sales to the Middle East is studied in Chichilnisky, 'The Role of Armaments Flows in the International Market', in D. A. Leurdijk and E. M. Borgesie (eds.), *Disarmament and Development* (RTO, Rotterdam, 1979). See also chapter of Chichilnisky and Heal, *The Evolving International Economy* (Cambridge University Press, 1987).
4. This figure is taken from J. Fabritius *et al.*., 'OPEC Respending and the Economic Impact of an Increase in the Price of Oil', in L. Matthiessen (ed.), *The Impact of Rising Oil Prices on the World Economy*, (The Macmillan Press, 1982).
5. These are set out in detail in Chichilnisky, 'Oil Prices, Industrial Prices and Outputs: A Simple General Equilibrium Analysis', and the related references in n. 2 above.
6. This estimate is given in G. E. J. Llewelwyn, 'Resource Prices and Macroeconomic Policies: Lessons from the Two Oil Price Shock', paper presented at the OPEC–UNITAR seminar on 'Oil and the North–South Agenda' (University of Essex, 1983).
7. Ibid.
8. These include Llewelwyn, op. cit., and William Nordhaus, 'Oil and Economic Performance in Industrial Countries', *Brookings Papers on Economic Activity*, 2 (1980), 341–99.

7

Oil as a Double-Edged Sword: the Development of Oil-Producing Countries

1. INTRODUCTION

When we think of oil-producing regions, we think of wealth. There is a stereotype of a rich Texan or a rich Arab, a stereotype which is closely associated with oil wealth. At this superficial level, few things seem more desirable than a large oil discovery. There are of course some figures which bear out these stereotypes: Fig. 7.1(*a*) and (*b*) show how the real capita income levels in selected oil-exporting countries rose as the price of oil rose after 1970. The figure also shows that by international standards, some oil-producing regions have achieved very high standards of living indeed.

In fact there is another aspect to the relationship between oil and wealth: many oil-producing regions have encountered serious economic difficulties, which have been associated with their oil-related activities.[1] Indeed, the issue is more general: exactly the same can be said of many countries which export other minerals.

One of the most widely noted cases is the effect on the Dutch economy of the exports of natural gas from the North Sea, an effect often referred to as the 'Dutch disease'.[2] Fig. 7.2 records the expansion of gas exports from the Netherlands, and also shows the movements in unemployment in Holland relative to the OECD average. These rose in parallel with the expansion of gas production. While these figures are not conclusive, they certainly suggest that there is something counter to conventional wisdom to be explained here.

A related phenomenon has been noted for Australia: a number of commentators have analysed the harmful effects of a

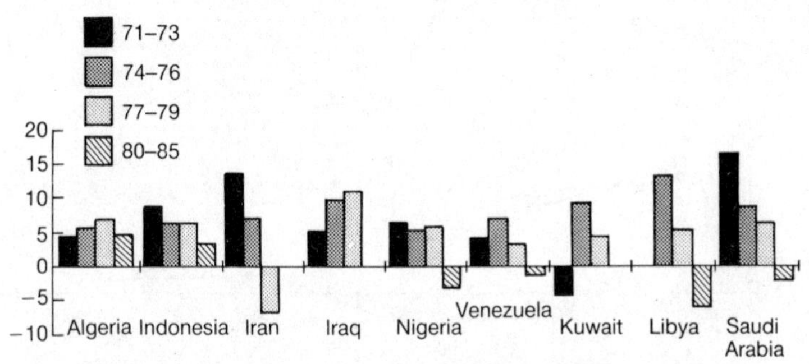

Fig. 7.1(a): Output growth rate for selected oil exporters.

Source: World Bank, *World Development Report*, 1987, and IMF (various statistics).

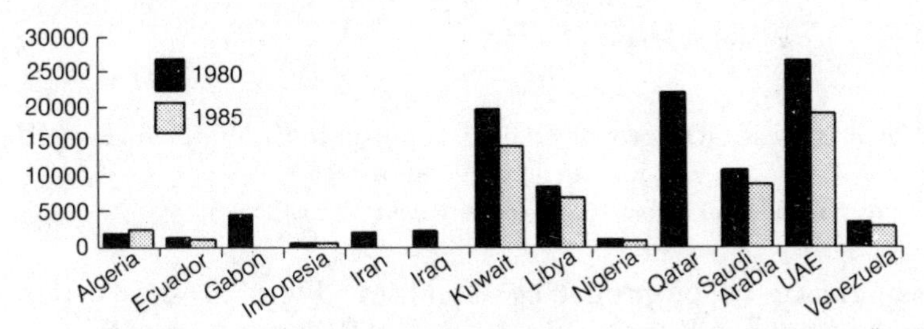

Fig. 7.1(b): GNP per capita (US$) for OPEC members.

Source: World Bank, *World Development Report*, (1982, 1987).

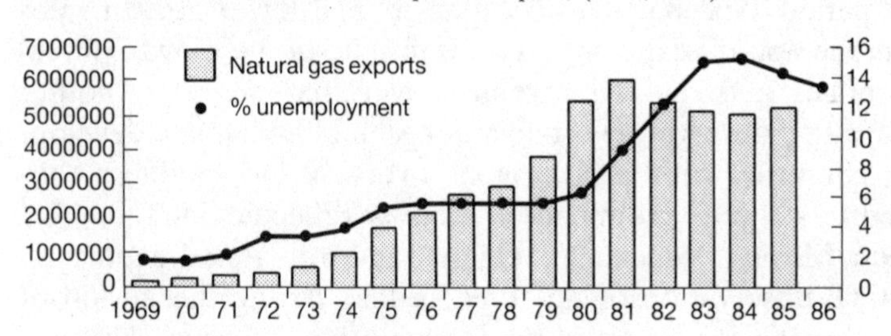

Fig. 7.2: Netherlands: gas exports US dollars (000's) and unemployment (%).

Source: *OECD Economic Outlook, UN Yearbook of Industrial Statistics* and *Yearbook of International Trade Statistics*.

boom in mineral exports on the agricultural and industrial sectors of the Australian economy.[3]

A more recent set of observations relates specifically to the

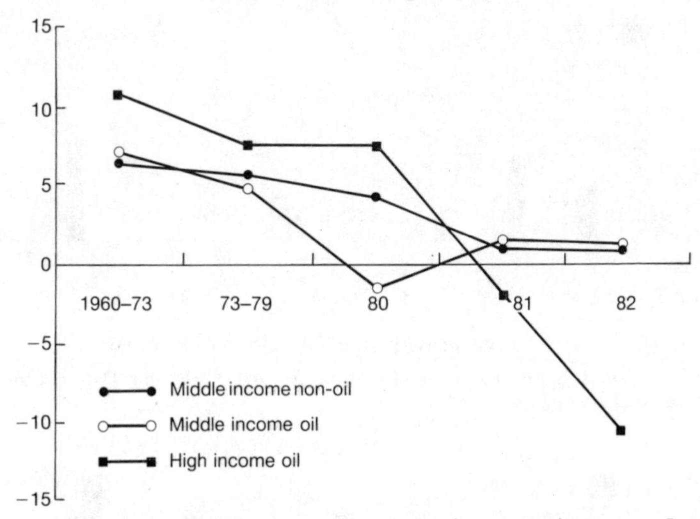

Fig. 7.3: Real GDP growth rates for oil exporters and for high- and low-income oil exporters.

Source: World Bank, *World Development Report* (various issues).

experience of oil-producing countries. Fig. 7.3 shows the growth rates of two classes of middle-income developing countries, those which export oil, and those which do not. Over the period 1960–83, the oil exporters had lower growth rates than the non-oil-exporters. This is in spite of the growth in real oil prices and real oil revenues over that period. Another interesting observation is that amongst middle-income developing countries with problems of overseas indebtedness, oil-exporters figure prominently. This is reflected in Table 7.1 where Mexico, Venezuela, Algeria, and Iran are oil-exporters. A final observation of this type is that during the period of rapid expansion of their oil sectors, Iran, Mexico, Nigeria, Indonesia, and Algeria all moved from being agriculturally self-sufficient to a position of substantial dependence on imports for meeting their domestic food requirements.

All of these examples suggest that oil production can create problems as well as solve them. It is clearly important to understand why this can happen, and to understand what determines the balance between harmful and beneficial results

Table 7.1: The twenty developing countries with the largest debt-service payments during recent years.

Country ranked by average debt service in 1980–1(*a*)	Debt service paid in US$ billions			
	1979*	1980*	1981*	1985†
1. Brazil	11.4	13.7	17.3	8.0
2. Mexico (*c*)	11.4	9.3	13.4	9.4
3. Venezuela (*b*)	2.8	4.7	6.0	—
4. Spain	3.0	3.7	5.0	—
5. Algeria (*b*)	3.2	3.9	4.4	1.3
6. Iran (*b*)	2.0	2.0	(6.1)	—
7. Yugoslavia	2.7	3.3	4.2	1.6
8. S. Korea	2.9	3.3	4.0	3.0
9. Argentina	2.1	2.8	3.7	—
10. Saudi Arabia (*b*)	2.9	3.1	3.4	—
11. Chile	1.7	2.2	3.1	1.6
12. Indonesia (*b*)	2.2	2.0	2.7	1.9
13. Egypt (*c*)	1.3	1.8	2.2	0.6
14. Peru (*c*)	1.1	1.6	2.0	0.3
15. Greece	1.1	1.3	1.7	—
16. Morocco	1.0	1.3	1.5	—
17. Nigeria (*b*)	0.8	1.2	1.6	1.3
18. India	1.1	1.4	1.4	1.1
19. Turkey	0.9	1.1	1.6	1.3
20. Philippines	1.3	1.1	1.6	1.0
Total 20 countries (*d*)	56.9	64.8	86.9	
Percentage of grand total LDCs	(75)	(75)	(80)	

(*a*) Next-ranking countries include United Arab Emirates, Portugal, Taiwan, Iraq, and Thailand. Debt-service payments by China PR in 1980 are tentatively estimated at $1.4 billion.
(*b*) OPEC Member.
(*c*) Net oil exporter.
(*d*) Total interest payments on long-term debt, in billion US$.
Source: * OECD, *External Debt of Developing Countries* (1982).
† *World Development Report* (1987).

from an expansion of the oil sector. It turns out that oil exports may have a substantial effect on the domestic economic equilibrium of the exporting country, and it is this effect which determines whether the eventual outcome is on balance favourable or unfavourable. The next step is therefore to analyse this effect.

2. THE EFFECT OF OIL EXPORTS

There are a number of mechanisms via which oil (or more generally mineral) exports may affect the exporting economy. We will discuss the simplest and most obvious first: this is the exchange-rate effect. An increase in foreign exchange earnings, at constant import levels, will lead to an appreciation of a country's exchange rate. This appreciation makes imports more attractive to its citizens, and makes its exports less competitive, or less profitable, on world markets. Hence its export and its import-competing sectors contract. This is exactly what is alleged to have happened as a consequence of Dutch gas exports. The Dutch exchange rate appreciated to a point where the manufacturing sector lost its competitive position in its traditional export markets. In fact, this phenomenon can occur even before mineral exports develop momentum. The mere discovery of large mineral deposits is sometimes enough to cause appreciation of a currency, as speculators anticipate the rise that will come later. And the inflow of foreign capital needed for the development of mineral resources may also cause an exchange appreciation by raising demand for the domestic currency, again before exporting actually begins.

It is important to see in detail what is happening here. An exchange rate appreciation raises the purchasing power of a country's residents on world markets, and so in principle increases their wealth and real income. However, this gain is spread unevenly: export and import-competing sectors lose, and others gain. In practice this may mean that the industrial and agricultural sectors lose, and these are of course just the sectors on which the majority of the population rely for their living.

The argument that resource exports raise the exchange rate at first sight seems convincing, so it is particularly important to realize that there is a need for qualification. The extra revenue from resource exports could in fact easily be absorbed by higher imports, or by overseas investment, without there being any significant exchange rate movement. Indeed it is a characteristic of many oil-exporting countries that imports and overseas investment have risen at least as fast as oil export revenues. The distributional impact of oil-related activities seems to encourage this type of response. If foreign-exchange-using activities such as importing or investing abroad respond sharply enough to an expansion of oil revenues, then this could neutralize fully the extra export revenues and leave the exchange rate unchanged. Indeed, there appear to be cases where this process has gone so far as actually to lower the exchange rate.

There are other mechanisms than the exchange rate, less obvious but possibly more important, through which resource exports can effect the internal economic configuration of the exporting country.[4] These are closely related to the income and substitution effects of oil price changes, effects which proved to be very important in discussing the macroeconomic impacts of changes in oil prices.

To understand these, let us begin by considering the consequences of an expansion of oil exports. With constant demand, the immediate consequence of an expansion of oil exports will be a drop in the nominal price of oil. If, as the empirical evidence suggests, the price elasticity of demand is less than one in absolute value, then this will lead to a fall in nominal oil revenues. So far, then, the oil producer is worse off. But matters will not stop there.

A drop in oil prices could lead to an economic expansion of some sectors of the consuming country, in effect shifting its demand curve for oil out to the right and compensating for some of the revenue lost from the drop in prices. There are further changes in the consuming country. This expansion together with the change in patterns of demands for inputs resulting from the change in oil prices, will change domestic factor prices, employment of factors and production costs, and so affect the prices of the consuming country's exports. This will alter the oil producer's terms of trade, the ratio of its export to import prices—which, as we have often observed before, are much more important to it than the nominal price of oil.

How are export prices likely to move? There is a cost reduction due to lower oil prices. In addition, factor prices in the oil-consuming country will tend to fall as oil prices fall and input demands shift to oil, lowering demand levels for factors competing with oil. On the other hand, an expansion induced by lower oil prices may tend to bid factor prices up. Overall, either outcome is possible, but for reasonable ranges of parameter values it seems likely that export prices will fall. In the oil-exporter's terms of trade, which equal the ratio of oil prices to industrial export prices, we therefore find that both the numerator and the denominator fall. The terms of trade will thus improve or worsen depending on whether the proportionate fall in industrial export prices exceeds or is less than the proportionate fall in oil prices. It is possible to determine precisely which proportional fall dominates: this answer depends on the details of the production structure of the industrial economy.[5] So, in summary, an expansion of oil exports may improve or worsen the terms of trade, depending on the characteristics of the oil-exporter's trading partner.

We still have to spell out the effect of higher oil exports on the internal workings of the exporting economy. Of course, if the terms of trade improve, this economy's real income rises. But even in this case, there are sector-specific effects that may be undesirable. For example, if the terms of trade improve, then real oil revenues rise. This leads to a higher level of domestic demand in the oil-producing country, and this demand deriving from oil revenues is likely to be focused primarily on industrial goods. Industrial goods are intensive in the use of capital rather than labour, so that an expansion in demand for these goods bids up capital income and lowers demand for labour. If this capital income is again spent primarily on industrial goods, then there is little or no gain to labour. The benefits from higher oil revenues are 'sealed into' the capital-intensive sectors, with little spillover to the wage-earning sectors. Indeed, it can be shown that real wages may actually fall.

If, on the other hand, the extra capital income is spent in significant part on labour-intensive goods, then the rise in oil-generated demand could spill over to the labour-intensive sector, but this theoretical possibility has seldom been observed in practice. So an expansion of oil revenues could benefit wage-earners either if the industrial sector uses both capital and labour on a significant scale, or if capital income is used to buy labour-intensive products. Otherwise, the benefits of higher real oil revenues may be concentrated entirely amongst capital-owners. Indeed, the realignment of demand towards a capital-intensive industrial sector can reduce the net demand for labour and lead to lower real wages. The basic point here is that whether wage-earners benefit from higher oil revenues depends on how integrated[6] the oil-producing economy is: a high level of integration, either in terms of both factors being used substantially in the industrial sector, or in terms of capital income generating demands for labour-intensive products, is needed to ensure that these benefits are diffused throughout the economy.

These arguments are intricate, and can only be appreciated fully in the mathematical model in which they were originally presented. For clarity, we summarize them as follows: an expansion of oil exports lowers *nominal* oil prices. This lowering of nominal oil prices may lead to either a rise or a fall in the oil exporter's terms of trade, the ratio of export to import prices. This is because the decrease in nominal oil prices can lead to a decrease in the prices of goods imported by the oil exporter, goods which use oil as an input. Therefore the denominator of the terms of trade goes down, as does the numerator, and the terms of trade themselves fall or rise depending on which falls faster, oil prices or the prices of goods traded in exchange for oil, such as industrial goods imported by the oil-producing country.

The oil exporter's terms of trade can therefore rise or fall: if they rise, domestic demand for industrial goods increases, leading to higher income for capital owners. Whether wage-earners also benefit, or whether they lose, depends on the degree of integration of the economy. In the worst possible case, then, the oil exporter suffers a drop in real wages as a result of an expansion of exports—though capital income rises.

This chain of reasoning is quite dependent on a set of assumptions about the structures of the two trading economies. How much industrial export prices fall, depends on the structure of the oil-consuming economy. This is the main determinant of the movement of the terms of trade. And the conclusion that real wages fall in the oil-producing economy depends on the assumption that there is little integration in this economy, so that the increase in oil-based demands stays within one sector. If, instead, the economy were rather integrated, then wages as well as capital income would rise as a result of higher oil exports. So the welfare consequences of an expansion of oil exports depend on the structural characteristics of the two trading economies.

An important result of Chichilnisky's paper, already hinted at above, is that an industrial economy is more likely than a developing economy to benefit from an expansion of oil exports by an increase in real wages. Industrial economies are typically more integrated and less 'dual'[7] than developing economies. In fact, the conditions most favourable to a beneficial outcome from an expansion of oil exports are likely to occur when an industrial economy (say Canada or the UK) is exporting oil to a developing country. Unfortunately, the conditions least likely to give a favourable outcome, are just those that are most widely observed: developing countries exporting oil to industrial countries.

The above arguments can readily be reversed. If an expansion of oil exports leads to lower wages, then a reduction of oil exports will raise wages. This might seem to have the implication that reducing oil exports to zero could be the best policy from the point of view of wage-earners. This is not so: there is an important non-linearity in the system, which reverses these effects at low export levels. So one can use this analysis to characterize a level of exports which is 'optimal' in terms of its impact on wages, or for that matter on any other variable. It is clear from what has been said that for different desiderata, these 'optima' need not coincide: the export policy that maximizes profit income may be quite different from that which maximizes real wages. But there may be a range where these goals converge, which could be useful for policy design.

As a final remark in this section, we want to emphasize that it is *not* being claimed that owning or discovering oil makes a country worse off. The claim is more specific and believable: it is that a rise in oil exports may harm particular sectors or groups in the exporting country and benefit others. The final outcome depends on the balance of the two sectors, and on the feedback that these sectors have on each other. Whether a positive or negative outcome emerges depends on the industrial structures of the trading partners. In general, industrial economies seem likely to be able to withstand the stresses induced by oil exports better than developing countries, because their economies are more homogeneous and better integrated.

3. CONCLUSIONS

Oil exporting can create economic problems as well as solving them. It seems likely that richer and more integrated economies will be better able to cope with these problems and benefit from oil exports. However, it should be noted that two of the countries most widely alleged to have suffered from resource exports, the Netherlands and Australia, are rich industrial countries. It is possible that the Australian economy is not integrated in the sense in which we are using this term, i.e. the feedbacks from the mineral sector to the remainder of the economy are small.

In the next chapter, we review in detail the available evidence about the effect of higher oil prices on economic development, both of oil-exporting and of oil-importing countries. With respect to oil-exporting countries, this evidence certainly suggests that the richer among them have on average coped with the strains of oil exporting better.

NOTES

1. For a catalogue of the problems that have beset the oil exporters, see, for example, Jahangir Amuzegar, 'Oil Wealth: A Very Mixed Blessing', *Foreign Affairs* (Spring, 1982), 814–36.
2. See, for example, W. M. Corden, 'The Exchange Rate, Monetary Policy and North Sea Oil: The Economic Theory of the Squeeze on Tradeables', *Oxford Economic Papers*, 33 (Special Issue, July 1981), 23–46.
3. See, for example, R. Gregory, 'Some Implications of the Growth of the Mining Sector', *Australian Journal of Agricultural Economics* (Aug. 1976), 71–91; and R. H. Snape, 'Effects of Mineral Development on the Economy', *Australian Journal of Agricultural Economics* (Dec, 1977), 147–56.
4. These arguments are set out in detail in Chichilnisky, 'International Trade in Resources: a General Equilibrium Analysis', *Proceedings of Symposia in Applied Mathematics*, American Mathematical Society 32 (1985), pp. 75–125. The rest of this chapter is essentially a non-technical exposition of the arguments in this paper.
5. The exact answer is given in Theorems 2 and 3 of Chichilnisky, op. cit. n. 13.
6. We used the word 'integrated' to describe an economy where the different sectors (industrial sector and traditional or labor intensive sector in developing countries) are clearly connected with each other in the sense that they use similar inputs (types of labour, capital, etc.) in similar proportions; they exchange goods and services with each other; and in general terms the economic fates of the different sectors are closely linked.
7. We use the word 'dual' to describe an economy which is not integrated. There are typically two sectors: the industrial sector and the traditional labour intensive sector. Each of these sectors uses different inputs (for example different types of labour) and has different factor proportions (for example the industrial sector is very capital intensive while the traditional sector is very labour-intensive). Finally, the sectors do not trade much with each other and their economic fates are not closely linked. The industrial sector may slump while the traditional sector expands and thrives, or vice versa.

8

Oil and the Developing Countries

In this chapter we will examine the impact of oil prices on the growth, trade, and debt of developing countries. The evidence shows interesting and unexpected differences between 'conventional wisdom' and the facts. We shall consider two periods: the period 1973–82 characterized by high oil prices and the period post 1982 during which oil prices were somewhat lower. In particular, this chapter shows that:

(*a*) Oil importing developing countries did not suffer a significant loss of growth or welfare due to higher oil prices;

(*b*) Growth rates of middle income oil *exporting* countries were lower than those of middle income oil *importers*;

(*c*) In the period of high oil prices the patterns of North–South trade and of South–South trade improved from the point of view of developing countries;[1]

(*d*) Other commodity prices moved initially in sympathy with oil prices and then dropped significantly while oil prices remained relatively stable;

(*e*) The oil-exporting developing countries have fared no better than the oil importers with respect to international debt.

A common line of enquiry links these issues: what is the impact of OPEC on the rest of the developing countries? Is their relationship one of co-operation, or of competition? Have OPEC actions been beneficial for the non-oil developing countries?

The period 1973–82

This period was characterized by high oil prices; it included the two major oil price 'shocks' of 1973 and 1979.

8.1 GROWTH OF THE OIL-IMPORTING COUNTRIES

Fig. 8.1 and Table 8.1 show that during the period 1973–82 the growth rates of the middle-income non-oil countries exceeded the growth rates of the oil-exporting middle-income countries: the first grew at an average rate of 4.48 per cent per annum between 1973 and 1982, and the second at the lower average rate of 3.48 per cent. Looking at the low-income developing countries, we see that their rates of growth averaged 4.85 per cent per annum over the same period.

It appears from this that the growth of non-oil *middle-income* countries was not seriously affected by the higher oil prices in the period 1973–82. Their rates of growth actually exceeded those of the oil-exporting middle-income countries over this period.

However, there could have been other adverse effects of oil prices on developing countries, and we focus on this possibility next. It is often argued that the cost of oil imports by low

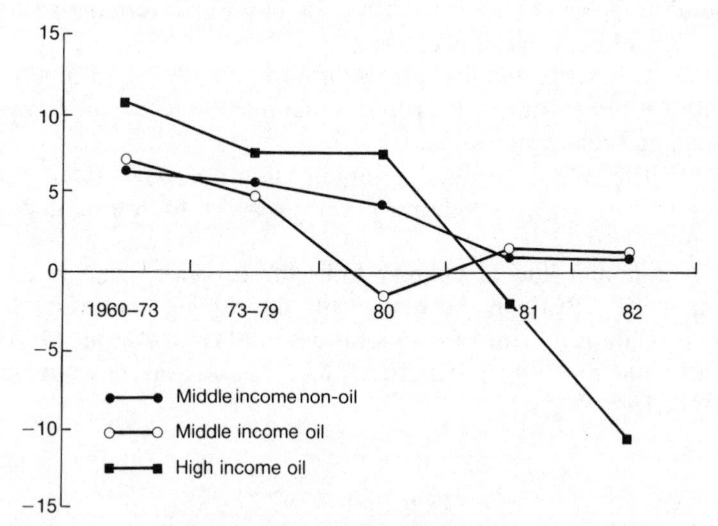

Fig. 8.1: Real GDP growth rates for oil exporters and for high- and low-income oil exporters.

Source: World Bank, *World Development Report* (various issues).

Table 8.1: Real GDP growth for industrial and developing countries, 1960–82.

	1960–73	1973–9	1980	1981	1982
Real GDP Growth					
Industrial countries	5.0	2.8	1.3	1.0	–0.2
All developing countries	6.0	5.1	3.0	2.0	1.9
Low-income countries	4.5	5.1	6.1	3.7	3
Non-oil-exporting middle-income countries	6.3	5.5	4.2	1.1	1
Oil-exporting middle-income countries	7.0	4.8	–1.3	1.5	1
High-income oil exporters	10.7	7.5	7.5	–1.8	–11

Source: World Bank, *World Development Report* (1984).

income countries increased significantly from 1973 to 1979, and that this produced proportionally more hardship on these economies than on the rest of the world. This argument sounds plausible, but what does the evidence disclose? Fig. 8.2 and Tables 8.2 and 8.3 examine the increases in the costs of oil imports of developing countries, as a percentage of their GDP, during the crucial period 1973 to 1980, when oil prices were highest. The value of imports of developing countries rose as a percentage of GDP during this period, confirming the view that higher oil prices were indeed a burden for these countries. But this burden and its rate of increase appear to be of a comparable magnitude to the burden that high oil prices inflicted on the North, as also shown by the data for Japan and the OECD in Fig. 8.2 and Tables 8.2 and 8.3. The percentage of oil imports in GDP of developing and of OECD countries increased over a comparable range. The explanation is simple: oil imports are a small percentage of GDP in the North, because their GDP is large; but they are also a small proportion of GDP in the South because their oil use is several times smaller than the OECD's. Fig. 8.3 shows that energy consumption per capita in low-income countries was about 421 kilograms of coal equivalent in 1979, while in the OECD it was 7,293 kg/ coal equivalent the same year.

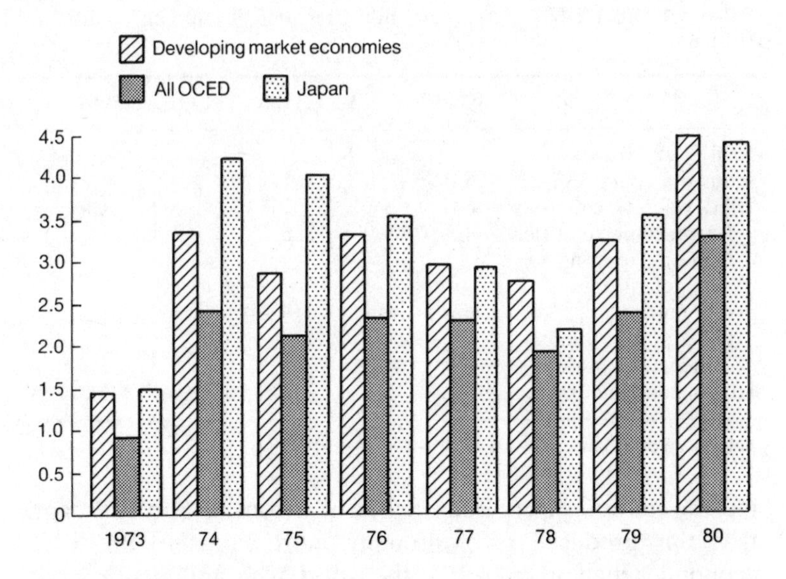

Fig. 8.2: Oil imports as a percentage of GDP for various country types.

Source: UN, *Yearbook of National Accounts*, World Bank, *Commodity Trade and Price Trends*, UN, *Yearbook of International Trade Statistics*.

Table 8.2: Oil Imports and GDP of developing market economies.

Developing market economies	GDP constant $ (bil.)	Crude imports constant $ (bil.)	Oil imports as percentage of GDP
1973	595.800	8.689	1.46
1974	825.500	27.698	3.36
1975	952.500	27.463	2.88
1976	1016.100	33.757	3.32
1977	1238.700	36.750	2.97
1978	1427.200	39.519	2.77
1979	1753.900	56.782	3.24
1980	2092.300	93.424	4.50

Sources: Country GDP: UN, *Yearbook of National Accounts*.
Crude Oil Imports: World Bank, *Commodity Trade and Price Trends*.

Table 8.3(*a*): Oil imports and GDP of the OECD countries.

Industrialized countries	GDP constant $ (bil.)	Crude imports constant $ (bil.)	Oil imports as percentage of GDP
1973	3254.600	30.438	.935
1974	3620.700	87.649	2.42
1975	4102.600	87.897	2.14
1976	4378.200	102.636	2.34
1977	5000.400	114.960	2.30
1978	6011.300	115.167	1.92
1979	6871.800	164.262	2.39
1980	7615.800	250.599	3.29

Sources: GDP: *UN Yearbook of National Accounts.*
Crude Oil Imports: World Bank, *Commodity Trade and Price Trends.*

Table 8.3(*b*): Oil imports and GDP, Japan.

	GDP constant $ (bil.)	Crude imports constant $ (bil.)	Oil imports as percentage of GDP
1973	379.64	5.686	1.50
1974	387.40	16.429	4.24
1975	385.66	15.624	4.05
1976	478.01	15.197	3.55
1977	548.79	16.056	2.93
1978	692.27	15.000	2.18
1979	559.22	19.895	3.56
1980	645.66	28.567	4.42

Sources: GDP: *UN Monthly Bulletin of Statistics.*
Oil Imports: *Yearbook of International Trade Statistics.*

Another significant element enters into this picture. This is the international solidarity of OPEC and other oil exporting countries in providing aid to the less developed countries during the period of high oil prices. Tables 8.4 and 8.5 present the empirical basis for this assertion. Table 8.4 shows that in the period 1975 to 1979 foreign aid transfers (ODA) from oil-exporting countries to low-income developing countries in Africa exceeded the cost of oil imports for the period: over the

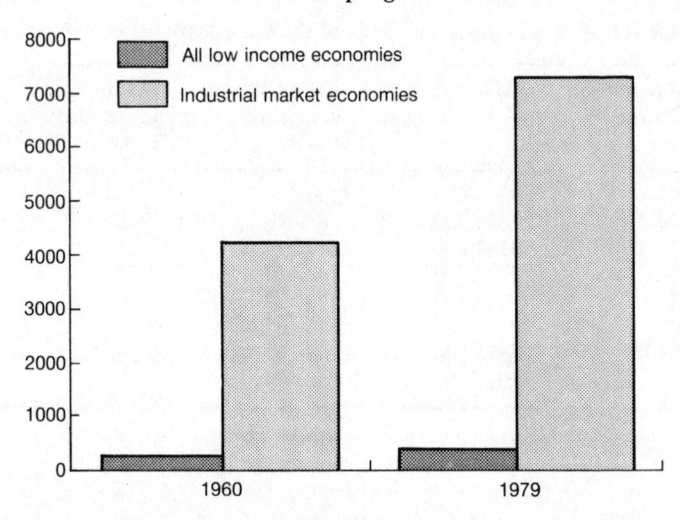

Fig. 8.3: Energy consumption per capita, kilograms of coal equivalent.

Source: World Bank, *World Development Report* (1987).

entire period 1975–81 the proportion of ODA to petroleum exports was 69 per cent. These transfers were in addition to a number of bilateral trade arrangements at preferential prices between oil exporters and less industrialized developing countries (LIDCs).

It is of interest to point out that in 1981 oil-exporting countries' transfers to LIDCs accounted for 1.4 per cent of their GDP (see Table 8.5), while during the same period OECD countries transferred less than 0.4 per cent of their GDP to low-income developing countries. By comparison, the UN target on aid is 0.7 per cent of GDP.

For the Gulf States and the USA, the figures are dramatically different: they transferred 3.85 per cent and 0.20 per cent of their GDP respectively as foreign aid in 1981. Indeed, in 1981 the Arab States were the world's largest aid donors, giving $7.6 billion per year, more than the USA and only slightly less than the entire European Economic Community.

In summary, the *low-income* developing countries were not

Table 8.4: Comparison of Gulf states' petroleum exports and bilateral Overseas Development Assistance (ODA) to non-petroleum-exporting African countries, 1975–81.

	1975	1976	1977	1978	1979	1980	1981
Gulf States' bilateral ODA to non-petroleum-producing countries	714.4	706.7	704.4	419.1	810.2	924.1	767.4
Gulf States' recorded petroleum exports to non-petroleum-producing countries	632.8	663.0	440.8	436.3	1012.9	1640.4	2480.8
Gulf States' bilateral ODA as percentage of petroleum exports to non-petroleum-producing African countries	112.9%	106.6%	158.9%	96.1%	80.0%	56.4%	30.9%

Weighted average of Gulf States' ODA to direct petroleum exports, 1975–81 = 69.0 per cent.
Sources: International Monetary Fund, *Direction of Trade Statistics, 1982* (Washington, DC, 1983), pp. 234–5, 317–18, 325–7, 375–6; and OECD, *Development Co-operation Annual Review* (Paris, OECD, various issues).

Table 8.5: Comparison of aid donors, 1981.

	ODA $ million	Share in world ODA (%)	ODA as percentage of GNP	Per capita income ($)
Arab Gulf States	7,317	20.5	3.85	16,120
Of which: Saudi Arabia	5,658	15.8	4.66	13,040
UAE	799	2.2	2.88	36,040
Kuwait	685	1.9	1.98	23,650
Qatar	175	0.5	2.64	26,520
Iraq	143	0.4	0.37	2,930
Libya	105	0.3	0.37	9,230
Algeria	65	0.2	0.16	2,120
Total Arab donors	7,630	21.4	2.55	6,230
Nigeria	149	0.4	0.17	1,000
Venezuela	67	0.2	0.10	n.a.
Iran	–150	–0.4	n.a.	n.a.
Total OPEC	7,696	21.5	1.40	n.a.
United States	5,783	16.2	0.20	n.a.
EEC	12,743	35.7	0.53	n.a.

Source: OECD, *Aid from OPEC Countries*, (1983), 15.

particularly harmed by higher oil prices. The growth rates of their oil import bills, as proportions of GNP, are comparable with those of the rest of the world, and they received very substantial aid flows from the oil producers in the period of high oil prices. In several cases the incremental OPEC aid flows exceeded the increased cost of oil imports. We have also seen that the growth of *middle-income* oil-importing developing countries was apparently not harmed by higher oil prices: actually their growth rates were higher than those of middle-income oil exporters during the period 1973–82. There is therefore no evidence of higher oil prices having had a proportionally larger adverse effect on developing countries, nor an overall adverse impact on the growth of oil-importing developing countries.

8.2 THE GROWTH OF OIL-EXPORTING DEVELOPING COUNTRIES

The oil-exporting developing countries fared in different ways according to their economic structures, as the previous chapter would lead us to expect. We have already seen that from 1973 to 1982 the middle-income oil countries grew less than the non-oil middle-income countries. However, the higher-income oil exporters fared rather differently: they recorded the highest growth rates over the earlier part of the period (7.4 per cent per annum from 1973 to 1980), and the lowest over the last part (−6.5 per cent per annum 1981–2) (see Fig. 8.1). Their mean growth rate over the period was almost identical to the non-oil middle-income countries and better than that of the oil exporting middle-income countries. So the high-income oil exporters had higher mean growth rates, but also had more variation in growth rates, than other oil exporters.

What explains this difference in GDP growth rates between high- and middle-income oil exporters? One hypothesis is that high-income oil exporters such as Saudi Arabia, United Arab Emirates, Kuwait, and Qatar have relatively simple economies where over 50 per cent of GDP is due to oil production. As this sector grew, the economies grew as well. The economy as a whole, therefore, followed the fate of the oil sector. The oil sector, in turn, followed the fate of oil prices. Thus, when oil prices were rising, the high-income oil exporters grew at very high rates; when oil prices stabilized or declined, they grew much less, or contracted. The growth rates of high-income oil exporters were therefore very sensitive to, and followed, the rates of growth of oil prices.

Middle-income oil exporters, such as Nigeria, Venezuela, or Mexico, behaved differently. These countries have more complex economies in which oil plays relatively a less important part than it does, for example, in the Gulf countries. Their growth is therefore less sensitive to oil price fluctuations, and indeed Fig. 8.1 shows graphically that the growth rates of middle-income oil exporters were much more stable over the 1970s than those of the high-income exporters.

The relative stability of growth rates in middle-income exporters is not only due to the fact that oil is a relatively less important source of income. These countries did not experience a great increase in growth rates during the periods of rapid price increases, and this is connected with some of the issues discussed in the last chapter: the impact of oil exports on exchange rates and on the rest of the domestic economy. There is no evidence that the exchange rates of these countries were unusually high, so it seems that the appropriate argument is the second one: the effect of oil exports on the internal economy of the exporting country. The expansion of oil sectors in these countries was accompanied by a decline in other domestic sectors, in particular in agriculture. Indeed, a striking aspect of Fig. 8.1 is that all oil-producing countries appear to have experienced a drop in their growth rates from the 1960s to the 1970s—even though oil prices were considerably higher in the 1970s.

8.3 NEW TRADE PATTERNS OF DEVELOPING COUNTRIES

The evolution of international trade by developing countries took a quantum leap in the period between 1973 and 1982. This is discussed in some detail in Chichilnisky and Heal (1987).[1] The developing countries became a much more important trading partner for the North, and indeed they represent at present approximately 40 per cent of the OECD export market. In 1970 the equivalent figure was only 27 per cent. The statistics show also that for the USA, EEC, and Japan, the developing countries are more important export markets than any of the two other developed partners together. This change in the role of developing countries in the world economy is clearly associated with the emergence of OPEC as a major purchaser in international markets. About one-third of the share of developing countries in OECD exports is explained by OPEC purchases.

The relative power of the partners in North–South trade therefore changed rather dramatically during the 1970s and 1980s. Since the main complaint about the organization of North–South trade has been that the North was disproportionally more powerful, this change indeed means that the distribution of power has moved in a more balanced direction over the period of high oil prices. This also gives a more solid basis to the idea of North–South interdependence: the North certainly now depends on the South for a significant share of its export markets.

These far-reaching changes in North–South trade were matched by important changes in South–South trade which are discussed in detail in Chichilnisky and Heal.[3] The statistics are clear: in the last decade, trade among developing countries was the most dynamic component of international trade, becoming in 1982 28.6 per cent of the share of developing countries' exports, and 7.6 per cent of world trade. Table 8.6 shows that in 1970 these figures were 19.6 per cent and 3.5 per cent, respectively. This figure reviews the growth in South–South trade as a proportion of the South's trade and world trade as a whole.

Within the rapid growth of South–South trade, manufactures were the most dynamic component, as shown in Table 8.8(b). By 1980 manufacturing represented almost 30 per cent of trade among developing countries. Developing countries' exports of

Table 8.6: The relative importance of developing countries' trade among themselves, 1970–82.

Year	Percentage share of developing countries mutual exports in their total exports*	Percentage share of developing countries mutual exports in total world exports*
1970	19.6	3.5
1971	20.1	3.5
1972	20.9	3.7
1973	22.0	4.0
1974	21.3	5.7
1975	24.6	5.9
1976	22.8	5.9
1977	23.8	6.1
1978	25.7	5.6
1979	24.3	6.2
1980	25.3	7.0
1981	26.4	7.3
1982	28.6	7.6

Sources: (1) Boris Cizelj, *Trade Among Developing Countries: Evaluation of Achievements and Potential* (Research Center for Cooperation with Developing Countries), p. 6, T. 1.
(2) UNCTAD Secretariat estimates, based on *United Nations Monthly Bulletin of Statistics* (June 1983 and earlier issues).
* Percentage of current dollar export values.

manufactures are less dependent now on industrial countries than are the other exports of these countries, a significant structural change. All this took place simultaneously with the rise in oil prices, a phenomenon which many authors associate with the structural changes in mutual trade among developing countries.

What is the link between higher oil prices and South–South trade? Oil countries became importers on a grand scale over the period of high oil prices and many of their imports were purchased from other developing countries. Table 8.7 shows this trend: OPEC imports from non-oil developing countries grew at an average rate of about 18 per cent from 1973 to 1980. In addition, oil-exporter's imports from other developing countries were different in nature from the imports of industrial

Table 8.7: Destination of non-oil-developing countries' exports and import volume growth in non-oil-developing countries' export markets, 1973–1980.

Volume of imports from developing countries (average percentage change)	1973	1974	1975	1976	1977	1978	1979	1980	Average 1973–80
Industrial countries	7.8	—	0.8	10.7	3.6	9.8	11.0	2.9	5.6
Oil-exporting countries	28.9	20.2	34.0	19.8	16.6	1.6	2.1	17.6	17.6
Non-oil-developing countries	27.5	12.8	2.1	6.9	5.3	12.4	15.2	16.2	11.8

Sources: IMF, *Direction of Trade*, various issues, and *World Economic Outlook* (various issues).

countries. Oil-exporting countries, many of which are not developed economies, imported technologically advanced manufactures and capital goods from other developing countries, sometimes as part of bilateral trade agreements. An interesting example is an arrangement under which the USSR and Turkey currently trade natural gas for consumer durables.

By contrast, industrial countries have traditionally imported labour-intensive manufactures and raw materials from developing countries, since the relative advantage of the industrial countries lies in their efficient production of technologically advanced and capital-intensive goods.

As a matter of fact, two major commodity groups made up most of the increase in mutual trade among developing countries: fuels and manufactured goods. Fuels rose from 37.3 to 47.1 per cent of mutual trade and manufactures from 15.8 to 26.9 per cent in the period 1955–79 (see Table 8.8(*b*)).

Table 8.8(*a*): Structure of developing countries'* mutual exports of manufactures 1973, 1975, 1977, 1979, 1980, in per cent.

Product category	1973	1975	1977	1979	1980
Non-ferrous metals	7.3	6.8	6.4	4.7	5.0
Iron and steel	5.9	5.1	4.7	6.7	5.7
Chemicals	12.6	12.3	10.5	10.5	11.2
Other semi-manufactures	10.2	10.2	9.9	10.9	10.9
Engineering products	30.3	36.7	38.2	39.0	39.2
Machinery for specialized industries	7.7	8.9	7.4	7.4	7.9
Office and telecommunications equipment	4.2	5.0	4.5	5.7	5.3
Road motor vehicles	3.6	4.6	3.6	3.8	4.0
Other machinery and transport equipment	10.2	13.6	16.6	15.7	14.6
Household appliances	4.5	4.6	6.1	6.4	7.4
Textiles	19.5	15.5	15.3	14.3	13.5
Clothing	6.0	5.4	6.2	5.0	5.6
Other consumer goods	8.3	7.7	8.8	9.0	9.0

* excluding traditional oil-exporting developing countries.
Source: Boris Cizelj, *Trade Among Developing Countries: Evaluation of Achievements and Potential*, and GATT, *International Trade* (1979/80, 1980/81).

Table 8.8(*b*): Structure of developing countries' mutual exports by major commodity groups 1955, 1965, 1975–9, in per cent.

Commodity group (SITC)	1955	1965	1975	1976	1977	1978	1979
Food (0+1+22+4)	27.6	27.3	15.5	12.7	13.4	13.5	12.7
Agricultural raw materials (2–22–27–28)	16.2	10.3	4.8	4.8	4.8	5.2	5.2
Ores and metal (27+28–67–68)	2.2	4.8	3.7	4.3	4.1	4.6	4.9
Fuels (3)	37.3	34.5	54.1	55.3	53.0	48.4	47.1
Manufactured foods (5 to 8 less 67+68)	15.8	21.6	21.3	22.4	24.2	28.0	26.9

Source: B. Cizelj, *Trade Among Developing Countries: Evaluation of Achievements and Potential*, and *Handbook of International Trade and Development Statistics* (1976, 1977, 1980, and 1981).

Certain major commodities decreased their share of South–South trade dramatically: food (from 27.6 per cent in 1955 to 12.7 per cent in 1979) and agricultural materials (from 16.2 per cent in 1955 to 5.2 per cent in 1979) as shown in Table 8.8(*b*). Over the period, developing countries increased their trade amongst themselves much more in fuels and manufactures, and much less in food and in agricultural raw materials. The other side of this coin in that developing countries have become increasingly dependent on food from industrial countries, and this is especially true for oil-exporting countries.

Table 8.8(*a*) taken together with Table 8.6 shows the changes in the structure of trade in manufactures among developing countries. Noteworthy is the growing share of engineering products (within manufacturing) which in 1980 accounted for almost 11 per cent of mutual exports (as opposed to 6 per cent in 1975).

8.4 COMMODITY PRICES

During the beginning of the decade, and following a period of expansion in demand, most commodity prices rose in sympathy with oil prices; this was true of such internationally traded commodities as copper, bauxite, coffee, etc. However, as the recession in the industrial countries set in, demand dropped and the prices of most commodities, except for oil, dropped as well. Fig. 8.4 illustrates these trends.

These movements of oil and commodity prices have been a source of great concern for oil-exporters and non-oil developing countries. The issue is whether the drop in the prices of other commodities was or was not 'caused' by the high prices of oil. A standard explanation is that high oil prices led to the recession in industrial countries and that this produced, with a lag, a drop in their imports of other commodities exported by developing countries. This presumably led to a drop in the prices of commodities other than oil.

However, the data discussed in Chapter 6 shows that higher oil prices could not be seen as the main 'cause', econometrically

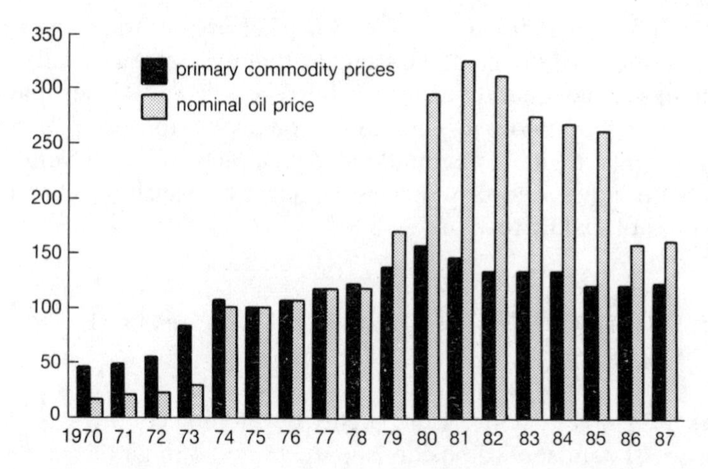

Fig. 8.4: Indices of non-oil primary commodity prices and of crude oil prices.

Source: *UN Monthly Bulletin of Statistics.*

or otherwise, of the recession in the industrial countries. Therefore high oil prices appear not to have 'caused' the drop in commodity demand and prices. Oil prices may be connected with other commodity prices, but this explanation seems flawed. Better explanations for current low commodity prices are required. These would include the level of interest rates, which are usually associated with changes in the prices of exhaustible resources, and other explanations of the cyclic behaviour of commodity prices.

The facts undeniably indicate a profound difference between the behaviour of oil prices and the prices of other commodities. Oil prices were sustained in the early 1980s, even in the face of relatively abundant supply and of slack demand. This is an indication of the relative market power of oil exporters, which derives, in economic terms, from the relative inelasticity of the demand for oil in the short term.

Other commodities mentioned here face a more price-elastic (and income-elastic) demand, and are sold in markets with a very different organization. This indicates that other commodities cannot follow the oil pricing policies of the 1970s unless there is a drastic change in market organization and in the elasticity of demand. This means that excessive specialization in such commodity exports is not a good idea. Rather than attempting to improve commodity prices or to reach price agreements in global negotiations, developing countries would be better advised to stop depending heavily on such unreliable sources of export revenues.

8.5 INDEBTEDNESS OF OIL IMPORTERS AND OIL EXPORTERS

Table 8.9 gives details of the twenty developing countries with the worst debt-service positions in the period of high oil prices. As already noted, oil-exporting countries feature prominently, with Mexico, Venezuela, and Algeria occupying three of the top five positions, and Iran and Saudi Arabia in the next five. Indeed, Argentina, which occupies the ninth position, is essentially self-sufficient in oil, so that only four of the ten most heavily indebted countries are oil importers.

Why are oil-exporting and oil-importing countries in such similar debt situations? An argument that has already been proposed is that oil-exporting did not yield the benefits that were widely expected, at least for middle-income countries. This point has been discussed at length, and does not require further elaboration. It is striking, however, that not only has the domestic growth of oil-exporting middle-income countries been relatively low, but also their balance of payments positions deteriorated during the oil-export expansion. Oil exporting was therefore not productive for these economies. In addition, oil exporting did not help these countries in an area which comes first to mind when recommending higher exports: the balance of payments.

Yet another connection exists between oil prices and the current debt problem. In the period 1973–1982 OPEC's export revenues from the two oil price rises (1973 and 1979) were largely reinvested through OECD banks and in particular in the Eurodollar market (see Table 8.10), and so increased the

Table 8.9: The twenty developing countries with the largest debt-service payments during recent years.

Country ranked by average debt service in 1980–1(*a*)	Debt service paid in US$ billions			
	1979*	1980*	1981*	1985†
1. Brazil	11.4	13.7	17.3	8.0
2. Mexico (*c*)	11.4	9.3	13.4	9.4
3. Venezuela (*b*)	2.8	4.7	6.0	—
4. Spain	3.0	3.7	5.0	—
5. Algeria (*b*)	3.2	3.9	4.4	1.3
6. Iran (*b*)	2.0	2.0	(6.1)	—
7. Yugoslavia	2.7	3.3	4.2	1.6
8. S. Korea	2.9	3.3	4.0	3.0
9. Argentina	2.1	2.8	3.7	—
10. Saudi Arabia (*b*)	2.9	3.1	3.4	—
11. Chile	1.7	2.2	3.1	1.6
12. Indonesia (*b*)	2.2	2.0	2.7	1.9
13. Egypt (*c*)	1.3	1.8	2.2	0.6
14. Peru (*c*)	1.1	1.6	2.0	0.3
15. Greece	1.1	1.3	1.7	—
16. Morocco	1.0	1.3	1.5	—
17. Nigeria (*b*)	0.8	1.2	1.6	1.3
18. India	1.1	1.4	1.4	1.1
19. Turkey	0.9	1.1	1.6	1.3
20. Philippines	1.3	1.1	1.6	1.0
Total twenty countries (d)	56.9	64.8	86.9	
Percentage of Grand Total LDCs	(75)	(75)	(80)	

(*a*) Next-ranking countries include United Arab Emirates, Portugal, Taiwan, Iraq, and Thailand. Debt-service payments by China PR in 1980 are tentatively estimated at $1.4 billion.
(*b*) OPEC Member.
(*c*) Net oil exporter.
(*d*) Total interest payments on long-term debt, in Billion US$.
Source: * OECD, *External Debt of Developing Countries* (1982).
† *World Development Report* (1987).

supply of loanable funds. This led to more borrowing. This point is further developed in Chichilnisky and Heal.[4]

Oil export revenues therefore provided liquidity to the international banking system during a period in which the

OECD countries were in a recession, and during which they enforced contractionary monetary policies.

At the end of the 1970s, however, things started to change. Interest rates in the USA increased threefold (from 6 to 18 per cent in the period 1976 to 1981) and the other OECD rates increased in sympathy, to avoid flight of internationally mobile capital. This sharply increased the burden of servicing the debt, much of which was in floating interest rate loans. Furthermore, during the early 1980s oil export revenues fell sharply, leading to a drop in deposits with OECD banks from the oil-exporters—the last columns of Table 8.10 show this clearly. There was therefore a double 'pinch' on the international financial system: a decrease in loanable funds and, simultaneously, much higher interest rates.

Interest rates in the USA remained at a historic high, so that dollar-denominated loans were a serious and threatening burden to the whole international banking system, for lenders as well as for borrowers. Furthermore, some of the most exposed borrowers were oil-exporting countries such as Mexico, Nigeria, Venezuela, and Ecuador. Some of these countries contracted their debts in order to develop their oil sectors, and indeed ended up exporting more oil, but at much lower prices. Their position became especially vulnerable.

The financial crisis also affected OECD countries indirectly. Oil-exporting countries purchased an important part of OECD exports: they borrowed to produce more and cheaper oil, and they mostly used the extra revenues to purchase goods from the OECD. When oil prices dropped, oil exporters decreased their imports from OECD. This affected OECD countries because these imports were an important addition to the lagging demand in OECD countries during the recession.

The willingness of oil exporters to export more oil and to import more industrial goods contributed to increasing the prices of OECD industrial goods and to lowering the price of oil. These are positive macroeconomic impacts which the oil countries had on the OECD, over and above the interest payments on the debt. Such gains must be taken into consideration both for understanding the origin of the debt

Table 8.10: Estimated deployment of OPEC countries' investment surplus, 1974–83.

	1974	1975	1976	1977	1978	1979	1980	1981	1982	1983
Identified investible surplus (*a*)	53.2	35.2	35.7	33.5	13.4	61.3	87.0	43.2	3.1	–19.6
Short-term investments	36.6	9.5	10.2	10.2	3.2	43.2	42.5	4.9	–16.2	–9.7
of which in:										
United States (*b*)	9.4	1.1	0.7	–0.5	–0.2	8.3	0.2	–3.5	4.8	–7.4
United Kingdom (*b*)	18.2	3.4	3.0	3.2	–1.6	16.2	16.1	7.9	–8.2	–5.1
of which in:										
Eurocurrency deposits	13.8	4.1	5.6	3.1	–2.0	14.8	14.8	8.1	–9.4	–5.5
in other industrial countries	9.0	5.0	6.5	7.5	5.0	18.7	26.2	0.5	–12.8	–8.8
Long-term investments	17.3	29.0	25.5	23.3	10.2	18.1	44.5	38.3	19.3	–3.8
of which in:										
United States	2.3	8.5	7.2	7.4	0.2	–1.5	14.3	15.3	7.6	–2.0
United Kingdom	2.8	0.9	1.4	0.6	–0.2	1.0	2.0	0.1	–0.8	–0.5
Other industrial countries (*c*)	3.1	5.8	4.3	5.8	2.6	8.7	16.7	13.6	6.6	–1.3

(*a*) The difference between the current-account position and identified foreign investment reflects, apart from recording errors, borrowing (net of repayments) by OPEC countries, direct investment inflows, trade credits and other unidentified capital flows.
(*b*) Including bank deposits and money-market placements.
(*c*) Bank deposits only.
(*d*) World Bank and IMF.
Source: *Bank of England Quarterly Bulletin* (June 1982).

problem and also for reaching constructive solutions to this problem. There is a new and powerful interdependence between developing and industrial countries to be taken into account.

8.6 CONCLUSIONS FOR THE HIGH OIL PRICE PERIOD

High oil prices have apparently not harmed the oil-importing developing countries in this period. Indeed, in the middle-income range, oil importers had higher growth rates than oil exporters. In the low-income range, aid and concessionary sales from OPEC have substantially offset the impact of rising oil prices.

Overall, the developing countries enjoyed relative prosperity in the 1973–82 period: growth rates were high and exports patterns improved. Some of these positive effects are attributable to high oil prices: investment was often financed by OPEC surpluses deposited in and lent on by OECD banks, and the booming OPEC markets boosted the exports of many other developing countries. Unlike a number of industrial countries, OPEC members have not raised discriminatory trade barriers against developing countries.

Some developing countries experienced serious problems, particularly with respect to their international trading and financial involvements. They are widely attributable to factors other than oil; one is the sharp rise in interest rates on their overseas borrowings which, as already mentioned, tripled in only four years. A second factor was the rise in the value of the US$ in the early 1980s: as most overseas borrowings are denominated in US$, this effectively raised the real value of debts outstanding. These two factors are related: the high value of the dollar is generally attributed to the high levels of interest rates in the USA. A third factor is the decline in the prices of the traditional exports of developing countries—primary products other than oil. These prices were at an all-time low in real terms at the end of the period, of course producing serious balance of payments problems for those countries dependent on their exports. Finally, as mentioned in Chapter 4, the prices of exports of industrial goods from the OECD countries to developing countries have risen sharply in the period 1972–82, cutting even further into the terms of trade of developing countries.

The period after 1983

By comparison with the 1973–82 period, the period from 1983 to 1990 was characterized by somewhat lower oil prices. There were no more oil price 'shocks' after 1979; OPEC's internal divisions increased and so did the supply of oil by both OPEC and by non-OPEC countries. Prices drifted lower since 1982 (see Fig. 8.4) and at the time this book goes to press real oil prices are close to those before the 1973 price increases.

So far this chapter has examined the impact of oil prices on growth, trade and debt of the developing countries, during the period of high oil prices, 1973–82. Our next task is to examine whether these conclusions change in the period post-1983, when oil prices were somewhat lower.

8.7 GROWTH OF OIL-IMPORTING AND OIL-EXPORTING COUNTRIES

During the 1983–6 period, middle-income oil importers grew more than oil exporters, the average rates of growth being 3.2 per cent and 0.37 per cent respectively. This is the same pattern as observed during the 1973–82 period: middle-income oil importers then also grew more than middle-income oil exporters. Table 8.11 provides the data. The main difference between the two periods arises in the growth rates of the high-income oil exporters. Their rate of growth from 1983 to 1986 was 1.3 per cent, while their growth during the period 1973 to 1982 was 6.4 per cent. In other words, during the period of high oil prices, high-income oil exporters did well; during the period of low oil prices, they did poorly. The growth pattern of high-income oil exporters thus follows the conventional wisdom: their fate is directly linked with the level of oil prices.

Table 8.11: Growth of real GDP, 1965–86 (annual percentage change).

	1965–73 avg.	1973–80 avg.	1981	1982	1983	1984	1985	1986
Developing countries	6.5	5.4	3.4	2.1	2.1	5.1	4.8	4.2
Low-income countries	5.5	4.6	4.8	5.6	7.7	8.9	9.1	6.5
Middle-income countries	7.0	5.7	2.8	0.8	0.0	3.6	2.8	3.2
Oil exporters	6.9	6.0	4.1	0.4	–1.9	2.3	2.2	–1.1
Exporters of manufactures	74.	6.0	3.3	4.2	4.9	7.8	7.8	7.0
Highly indebted countries	6.9	5.4	0.9	–0.5	–3.2	2.0	3.1	2.5
Sub-Saharan Africa	6.4	3.2	–1.0	–0.2	–1.5	–1.7	2.2	0.5
High-income oil exporters	8.3	7.9	1.4	–0.5	–6.9	1.2	–3.8	8.2
Industrial market economies	4.7	2.8	1.9	–0.5	2.2	4.6	2.8	2.4

Source: World Bank, *World Development Report* (1987).
Note: Data for developing countries are based on a sample of ninety countries. Data for 1986 are estimates.

Middle-income oil exporters, on the other hand, exhibit less conventional behaviour. Their fate is not so closely linked with oil prices: with high oil prices these countries grew less than their oil-importing counterparts; with low oil prices they grew less too. However, their rates of growth were higher in the period of high oil prices. The previous section analysed the reasons for these growth patterns, which appear to hold in the post-1983 period as well.

8.8 TRADE, DEBT, AND AID

Trade patterns among developing countries changed significantly in the period of high oil prices. They were examined in section 3 above.[5] The debt burden of the oil-exporting countries showed little difference from that of oil importers in the 1980–4 period. Table 8.12 shows that this was both in terms of debt service to export ratios and in terms of debt service to GNP ratios. In the post-1983 period the main developing countries' debtors are still oil-exporting countries, as was shown in Table 8.9.

The average ratio of debt service to GNP during 1980–4 for energy importers was 4.4 per cent; the same average for energy exporters was 4.7 per cent. The energy exporters therefore show a higher debt service burden over this period. As shown in Table 8.13, during the 1981–4 period, the ratio of interest payment to GNP was 2.35 per cent for energy importers and 2.39 per cent for energy exporters. During this period therefore the debt burden, measured by ratios of debt service and interest payments to GNP, was somewhat higher for energy exporters than for energy importers, again confirming the observation that oil, or energy, exports cannot be expected to improve the debt position of developing countries.

Finally, we recall that OPEC solidarity in the form of aid, greatly decreased the developing countries' burden from increased oil prices, more than compensating in the case of the African nations. In the post-1983 period, the OAPEC countries remained the largest aid donors as a proportion of GNP: their

Table 8.12: Impact of changes in terms of trade and debt burdens.

	Impact of terms of trade on GDP, 1981–4* (Annual rates per cent) (*a*)	Debt/GNP ratio			Debt/export ratio			Debt service/ GNP ratio (av. 1980 –4) (per cent of GNP)	Debt service/ export ratio (av. 1980 –4) (per cent of exports
		1970	1980	1984	1970	1980	1984		
Fast-growing countries									
Net energy importers									
Republic of Korea	0.0	22.4	29.5	37.0	143.0	80.1	89.0	6.0	15.0
Pakistan	–0.4	30.6	34.4	29.7	380.7	266.5	258.7	2.4	21.0
Singapore (*b*)	0.3	7.9	12.1	10.6	7.6	5.3	6.0	2.0	1.0
Hong Kong (*b*)	2.2	0.1	1.8	0.9	0.1	1.8	0.8	0.3	0.3(d)
Burma	–0.2	4.7	25.2	3.9	77.9	293.7	517.4	2.3	27.7
Thailand	–1.1	11.1	17.7	26.3	62.0	70.7	114.3	4.3	18.3
Sri Lanka	0.7	16.1	33.7	41.9	83.7	100.6	137.1	2.8	9.4
India	0.1	15.1	11.1	13.7	356.5	145.3	173.5	0.9	11.8
Turkey	–0.4	14.8	27.0	32.3	244.4	428.4	166.1	3.4	25.6
Mean	0.1	16.1	18.3	21.9	182.9	92.2	94.0(e)	2.3	11.4(e)

Net energy exporters									
Oman	—	—	8.2	17.2	—	11.7	26.6	2.4	3.6
Cameroon	–1.2	13.0	33.6	31.3	50.3	106.1	94.0	4.2	13.6
Congo	0.6	53.9	71.4	76.2	179.6	110.0	—	8.7	14.2(d)
Malaysia	–1.0	10.0	16.8	39.4	21.3	26.1	62.1	3.0	5.0
Mean	–0.5	10.2	18.6	34.5	23.0	31.5	58.9	3.2	5.6(f)
Mean of other developing countries									
Net energy importers	–0.4	18.6	27.5	50.3	102.7	110.8	169.6(c)	6.5	22.6(c)
Net energy exporters	0.1	18.9	24.7	41.6	130.3	93.4	176.7	6.3	25.8

(*a*) Defined as percentage change in terms of trade multiplied by the ratio of exports to GDP.
(*b*) Debt ratios cover only public and publicly guaranteed debt.
(*c*) 1983.
(*d*) Excluding Hong Kong.
(*f*) Excluding Congo.
Note: The means, except for terms of trade, are weighted averages.
Source: World Economic Survey, UN (1987).
Source: World Bank, *World Development Reports*, and *World Debt Tables*, 1985–1986.

Table 8.13: The relative interest payment burden of the fast-growing and other countries.

	Ratio of debt-service to GNP (av. 1981–4) (per cent of GNP)	Ratio of interest payment to GNP (av. 1981–4) (per cent of GNP)
(a) Average for fast-growing energy importers	2.30	1.07
(*b*) Average for other energy importers	6.5	3.65
(*b*)/(*a*)	2.83	3.41
(*c*) Average for fast-growing energy exporters	3.15	1.83
(*d*) Average for other energy exporters	6.25	2.95
(*d*)/(*c*)	1.98	1.61

Source: UN *World Economic Survey* (1987).

contribution represented 1.6 per cent of their GNP in 1985, with the USA's proportion being 0.23 and the OECD's 0.36 per cent (see Table 8.14). However, in absolute value, the total contributions of the OECD countries were higher in 1985 than those of the OPEC countries due to the fall in the latter's income with the lower oil prices.

NOTES

1. South denotes the developing countries, North the industrial countries. Planned economies are excluded.
2. Chichilnisky and Heal, *The Evolving International Economy* (Cambridge University Press, 1987).
3. Ibid.
4. Ibid.
5. There are opposing views on the trends in South–South trade. Havrylyshyn and Wolf argue that 'contrary to widespread impression,

Table 8.14: Comparison of aid donors, 1985.

	ODA ($ million)	ODA as per cent of GNP	Per capita income ($)
Arab Gulf States			
Saudi Arabia	2,646	2.88	8,850
UAE	58	0.24	19.270
Kuwait	749	3.16	14,480
Qatar	–2	–0.03	n.a.
Iraq	–26	–0.08	n.a.
Libya	151	0.49	7,170
Algeria	45	0.08	2,550
Total OAPEC	3,621	1.6	
Nigeria	45	0.17	800
Venezuela	32	0.10	3,080
Iran	–171	–0.11	n.a.
Total OPEC	3,527	1.06	
USA	9,784	0.23	16,690
OECD	37,060	0.36	11,810

Sources: *World Development Report* (1987).

there has been no large shift towards trade among developing countries since 1963, at least when the focus is upon non-fuel trade and developing countries exclude capital-surplus oil-exporting countries'. See 'Recent trends in trade among developing countries', *European Economic Review*, 21 (1983), 333–62. Yet the same authors agree that the share of the South (OPEC plus other developing countries) as a recipient of total Southern exports rose from 12.8 per cent in 1973 to 28.5 per cent in 1980 (and further to 30.3 per cent in 1981). As for manufacturing exports from the South S. Lall establishes in *Trade and Development, UNCTAD Review* No. 6 (1985), that 'they were 28.5% in 1973, 36.2% in 1980 and 37.3 % in 1981. Of the 7.7% points increase in the share of the South in 1973–80, the OPEC countries accounted only for 3.8 points. In terms of shares the oil importing countries as a group displaced OPEC as the most dynamic market for the South's manufactured products.'

9

Summary

9.1 OIL PRICE MOVEMENTS

The previous chapters have reviewed salient micro- and macroeconomic themes in an attempt to understand the international oil market and its position in the world economy. A number of important factors have emerged.

One is the trend of price dynamics in this market. Growing scarcity is an underlying theme in the long run, an underpinning to the market, giving an upward trend to prices. But this long-term trend can be interrupted by discoveries, by demand shifts resulting from macroeconomic policies, by interest rate movements, by technological changes, or by changes in the market power of the dominant producers.

Another important issue is the big difference between the long- and short-run responses to a price change, on both demand and supply sides. In the short run these responses are minimal; in the long run (which may mean a decade or more) they are very large. Such long lags between a cause and its full effects are in general very destabilizing. They induce agents to pursue policies to extremes which are currently sustainable, but which in the longer term produce strong adverse effects. These adverse effects lead to strong evasive measures which in turn have their own adverse long-run consequences. The oil market may be moving into such cycles.

Overall, then, the picture that emerges of price dynamics in the oil market is a complex one. In spite of this, it is possible to give a relatively simple four-stage summary of what happened in the world oil market from 1970 to 1989:

(*a*) A period of low prices led to inelastic demand, to high consumption levels and to a decline in the importance of non-OPEC output. This set the stage for:

(*b*) A period when OPEC was able to assert market power

through a combination of economic circumstances deriving from (*a*) and also political circumstances in terms of Arab–Israeli relations.

(*c*) The sharp price increases of (*b*) were followed by very predictable, though relatively slow, demand and supply responses. Non-OPEC output rose, demand both fell and became more elastic, and prices were sustained only by significant reductions to OPEC output.

(*d*) Finally, the downward pressure on prices from the increases of the 1970s and the slow responses in (*c*) became too great for OPEC to counteract and prices dropped sharply, almost back to their starting values in real terms. By the end of the cycle, OPEC's role in the market had been considerably reduced, perhaps back to that which it played in the late 1960s and early 1970s.

Looking ahead, there is a good chance that at least the earlier stages of this cycle could be repeated. Indeed, if technological advances make large-scale production from unconventional sources possible at competitive prices, the whole cycle may be repeated, with the price rises leading to increased non-OPEC production from unconventional sources.

The skeletal outline above has to be fleshed out by considering the speculative element in commodity markets—the tendency of price movements to become self-reinforcing—and also by considering the effect of interest rates on price movements. As noted in Chapter 5 (and in Fig. 5.2), major price rises have coincided with falling interest rates and vice versa.

Overall, it would be reasonable to characterize the developments of the last two decades by saying that oil has become 'more like other commodities': its price has become more subject to short-term market forces, the market has become more speculative, and has moved away from the earlier regimes where prices were administered by the major oil companies, or by OPEC, or by combinations of the two. The difference between long- and short-term elasticities, and the importance of speculative elements in the market, are important in many commodity markets and are major contributors to the price volatility which characterizes these markets.

On the issue of OPEC's role in the world oil market, the verdict must be that OPEC, while a major player, is not able to determine the level of oil prices in the long run. They face many countervailing forces: market responses on the supply and demand sides, though slow, are strong, and there are many other influential players in the market. It is also clear that OPEC's role in the world oil market has not had a major negative impact on oil-importing developing countries.

9.2 OIL AND THE MACRO-ECONOMY

There are also many interesting issues in the relationship between oil prices and the macroeconomic environment. The first point to note here is that causal relationships are not one way, only from oil prices to macro-variables such as inflation and unemployment. Oil prices and their movements can clearly be affected by macro-variables such as interest rates and the level of aggregate demand.

If we look at the effects of oil prices on such macro-variables as inflation, unemployment, and productivity, it is clear that high oil prices on balance had a negative impact, though probably rather limited in magnitude. This negative effect was the net effect of an impact which was positive in some sectors and negative in others, depending on whether they were substitutional or complementary with respect to oil use. The effect of the decades of the 1970s and 1980s on the oil producers themselves is also interesting. Some of the Gulf states have become very rich, but many of the other oil producers have gained much less than might have been anticipated. This is a phenomenon which is related to their internal economic structures and the extent to which those provide linkages between sectors. Their experience shows clearly that a country's ability to benefit from oil exports depends crucially on its internal economic structure and on the policies which its government adopts. Oil exports lead to losses rather than gains for many oil exporters.

9.3 OIL AND INTERNATIONAL FINANCIAL MARKETS

There are two aspects of oil markets which we have chosen not to study: these are their impact on the international financial system, and also their environmental dimension.

Changes in financial flows associated with the changes in oil prices have had an impact on the development of international financial markets. Table 9.1 shows that in 1974–7, and again from 1979 to 1981, the major oil exporters had large current account surpluses. Many of these surpluses were initially deposited with Western—and primarily European—financial institutions on a short-term basis, while their owners made decisions about their longer-term deployment see Table 8.10 for more detail. The financial institutions receiving these deposits naturally wanted to find profitable ways of lending them on, and in a period characterized by recession or slow growth in the industrial countries but by relatively rapid growth in many developing countries (as shown in Fig. 8.1), the receiving banks used the inflow of OPEC surpluses to make unusually large loans to many developing countries. These

Table 9.1: Current account balances of major oil exporters (in US$ billions).

Year	Major oil exporters
1973	6.7
1974	68.3
1975	35.4
1976	40.3
1977	29.4
1978	–1.3
1979	56.8
1980	102.4
1981	45.8
1982	–17.8
1983	–18.0
1984	–10.0
1985	–5.5

Source: International Monetary Fund, *World Economic Outlook*.

loans were typically made at floating interest rates, and on a relatively short-term basis. The sharp rise in real interest rates in the early 1980s, coupled with the maturity of many of the initial loans, then produced the Third World debt crisis which has been a prominent feature of the international economic scene ever since, and which is having a major long-term impact on the international financial system. So the debt crisis of the 1980s has its origins in the OPEC surpluses of the 1970s—though this is not to say that OPEC is in any way responsible for it: the lending policies of the banks, the borrowing policies of their customers, and the monetary policies of the USA and the UK completed a process which OPEC started.

The OPEC surpluses of the 1970s also contributed to the rapid growth of the Eurodollar security market in the 1970s: this provided an unregulated, anonymous, and relatively liquid market where OPEC members could make short-term investments which they felt were safe from any form of retaliation by the American or British governments. In the longer run, many of these funds went into equity and property investments in Europe and the USA, and contributed part of the initial impetus to the development of the cross-border equity market.

9.4 ENVIRONMENTAL ISSUES

Environmental problems are associated with most widely used sources of energy—coal, nuclear power and oil. Of these energy sources, oil is perhaps the 'cleanest', being associated only with the production of controllable amount of nitrous oxide and sulphur dioxide, and with the emission of carbon dioxide, which is of course an unavoidable by-product of the use of any fossil fuel.

Until recently the production of carbon dioxide was not widely recognized as environmentally harmful. Within the last few years, however, it has been recognized that the emission of carbon dioxide from the combustion of fossil fuels is leading to a 'greenhouse effect', which is slowly changing the earth's climate and weather patterns. While there is still uncertainty about the nature and extent of these changes, there does seem to be a significant risk that they will be harmful, substantially reducing the productivity of many currently important agricultural regions.

The problem is a hard one to tackle, because it is truly global in its dimensions. The issue is global climate change, caused by the release of carbon dioxide anywhere in the world. It can therefore only be fully resolved by a co-ordinated international move to control the use of fossil fuels, something which will not occur easily. Nevertheless, this is likely to be on the political agenda of the 1990s, and may lead to restrictions on the use of oil. Of the current widely used power sources, only nuclear power does not produce carbon dioxide. If the other risks associated with nuclear power can be controlled, then concern with the greenhouse effect may lead to a regulatory environment favouring nuclear power at the expense of fossil fuels, at least in the firing of power stations, a field where coal, oil and nuclear are all usable. Oil seems likely to remain the dominant fuel source in transportation for a very long time.

Bibliography

M. A. ADELMAN *The World Petroleum Market* (Johns Hopkins University Press, 1972).

TERENCE AGBEYEGBE 'Interest rates and metal price movements', *Journal of Environmental Economics and Management* (forthcoming).

JAHANGIR AMUZEGAR 'Oil wealth: a very mixed blessing', *Foreign Affairs* (Spring, 1982), 814–36.

E. BERND and D. W. WOOD 'Technology, prices and the derived demand for energy', *Review of Economics and Statistics*, 56, (1975) 259–68.

E. BERND and D. E. WOOD 'Engineering and econometric interpretations of energy-capital complementarity', *American Economic Review*, 69, (1979) 342–54.

British Petroleum *Annual Review of Energy* London, various issues.

GRACIELA CHICHILNISKY 'Oil prices, industrial prices and outputs: a simple general equilibrium analysis' (discussion paper, Economics Department, Columbia University, 1980).

—— 'A general equilibrium theory of north–south trade', ch. 1, vol. II pp. 3–56 in Heller, Starr, and Starret (eds.) *Essays in the Honor of Kenneth J. Arrow* (Cambridge University Press, 1988).

—— 'International trade in resources: a general equilibrium analysis', in R. McKelvey (ed.) *Environmental and Natural Resource Mathematics* (Proceedings of Symposia on Applied Mathematics, American Mathematical Society, 1985) 75–125.

—— 'Prix du petrole, prix industriels et production: une analyse macroeconomique d'equilibre general in G. Gaudet and P. Laserre (eds.) *Ressources Naturelles et Theorie Economique* (Quebec, Les Presses de l'Universite de Laval, 1986), pp. 26–56.

—— *Necessidades Basicas, Recursos, Naturales y Crecimiento en el Contexto Norte-sur:* Respuesta a un Comentario (Desarollo conomico, 1986), 97, vol. 25, (1985), pp. 128–33.

—— 'Resources and north–south trade: a macro analysis in open economies' (discussion Paper, Columbia University Department of Economics, 1981), published in H. Singer, N. Hatti and R. Tandon (eds.) *Challenges of South–South Cooperation* (Ashish Publishing, New Delhi, 1988).

—— 'The role of armament flows in the international market', in D. A. Leurdijk and E. M. Borgesia (eds.), *Disarmament and Development* (RIO, Rotterdam, 1979).

—— 'Necesidades basicas, recursos no renovables y crecimiento en el contexto de las relaciones norte-sur, *Desarrollo Economico*, no. 94, vol. 24 (July–September 1984), pp. 171–86.

—— 'Oil prices and the developing countries—the evidence of the last decade', Intereconomics, vol. 20 no. 6 (December 1985), 288–95.

—— and Geoffrey Heal *The Evolving International Economy* (Cambridge University Press, Cambridge, 1987).

—— —— 'Energy–Capital substitution: a general equilibrium analysis', Collaborative Paper, IIASA, Laxenburg, Austria, 1983).

—— —— and Darryl McLeod, 'Resources naturelles, commerce et endettement'. In G. Gaudet and P. Laserre (eds), 'Resources naturelles et theorie economique', Quebec, Les Presses de l'Universite Laval, 1986, 57–90.

—— —— *Resources, Trade and Debt: the Case of Mexico* (The World Bank, Global Analysis and Projections Division, Division Working Paper No. 1984–5).

—— —— Geoffrey Heal and Amir Sepahban, 'Non-conflicting oil pricing policies in the long run', *OPEC Review* (1983), 330–56.

—— A. Hererra, H. Scolnick, *et al.*, *Catastrophe or New Society* (International Development Research Center, Ottowa, 1976).

—— 'Trade and development in the 1980s', ch. 18, pp. 195–240, in Ashok Bapna (ed.), 'One world one future: new international strategies for development' (Preger Publishers, New York, 1986).

Boris Cizelj *'Trade Among Developing Countries: Evaluation of Achievements and Potential* (Research Centre for Cooperation with Developing Countries, 1982).

W. M. Corden 'The exchange rate, monetary policy and north sea oil: the economic theory of the squeeze on tradeables', *Oxford Economic Papers*, 33 (1981), 23–46.

Partha Dasgupta and Geoffrey Heal 'Economic theory and exhaustible resources'. Cambridge University Press, Cambridge, 1979.

M. Denny, M. A. Fuss and L. Waverman 'The substitution possibilities for energy: evidence from US and Canadian manufacturing industries' (Working paper No. 8013, Institute for Policy Analysis, Toronto, 1980).

Energy Modelling Forum Working Group, 'Aggregate elasticity in energy demand', *Energy Journal*, vol. 2, no. 2, (1981), 37–75.

J. Fabritius *et al.*, 'OPEC respending and the economic impact of an increase in the price of oil', in L. Mattheisen (ed.) *The Impact of Rising Oil Prices on the World Economy*, (Macmillan Press, 1982).

J. W. Forester *World dynamics* (Wright Allen Press, 1971).

Dermot Gately, 'Do oil markets work? Are OPEC dead?' *Economic Research Report* (C. V. Starr Center for Applied Economics, New York University, 1988). Published in *Annual Review of Energy*, vol. 14.

L. C. GRAY 'Rent under the assumption of exhaustibility', *Quarterly Journal of Economics* (May 1914).

R. GREGORY 'Some implications of the growth of the mining sector', *Australian Journal of Agricultural Economics*, (1976), 71–91.

JAMES B. GRIFFIN and HENRY B. STEELE Energy economics and policy (Academic Press, 1986).

J. M. GRIFFIN and E. R. GREGORY 'An intercountry translog model of energy substitution responses', *American Economic Review*, 66 (1976) 845–57.

D. J. TEECE *OPEC Behavior and World Oil Prices* (George Allen and Unwin, 1982).

JOHN M. HARTWICK and NANCY D. OLEWILER *The Economics of Natural Resource Use* (Harper and Row, 1986).

O. HAVRYLYSHYN and M. WOLF 'Recent trends in trade among developing countries', *European Economic Review*, 21 (1983), 333–62.

GEOFFREY HEAL 'Uncertainty and the optimal supply policy for an exhaustible resource', in R. S. Pindyck (ed.), *Advances in the Economics and Energy Resources* vol. 2 (JAI Press, 1979), 119–47.

—— 'Review of books on energy demand', *Economica* (1979), 322–3.

—— and MICHAEL BARROW 'The relationship between interest rates and metal price movements', *Review of Economic Studies*, xlv.11, (1980), 161–81.

—— —— 'Empirical investigation of the long-term movements of resource prices: a preliminary report', *Economics Letters* 7. (1981) 95–103.

WILLIAM W. HOGAN 'Patterns of energy use revisited' (Discussion Paper, Energy and Environmental Policy Center, John F. Kennedy School of Government, Harvard University, June 1988).

HAROLD HOTELLING 'The economics of exhaustible resources', *Journal of Political Economy* 39 (1931), 137–75.

S. LALL 'Trade and development', *UNCTAD Review* (1985) no. 6.

G. E. J. LLEWELWYN, 'Resource prices and macro-economic policies: lessons from the two oil price shocks'. Paper presented at the OPEC-UNITAR Seminar on 'Oil and the north-south agenda' (University of Essex, 1983).

S. L. MCDONALD *Petroleum Conservation in the United States: an Economic Analysis* (Published for Resources for the Future In. by the Johns Hopkins University Press, 1971).

G. S. MADDALA, W. S. CHERN and G. S. GILL *Econometric Studies in Energy Demand and Supply* (Praeger Publishers, New York and London, 1978).

D. H. MEADOWS *et al. The Limits to Growth* (University Books, New York and Earth Island Press: London, 1972).

MERTON MILLER and CHARLES UPTON, 'A test of the Hotelling valuation principle', *Journal of Political Economy*, 93 (1985), 1–25.

WILLIAM D. NORDHAUS, 'The allocation of energy resources', *Brookings Papers on Economic Activity*, 3 (1973), 529–76.

—— 'International studies of the demand for energy', North Holland, 1977.

—— *The Efficient Use of Energy Resources* (Cowles Foundation Monograph, Yale University Press: New Haven and London, 1979).

—— 'Oil and economic preference in industrial countries', *Brookings Papers on Economic Activity*, 2, (1980), 341–99.

OECD *Inflation—the Present Problem* (Paris, 1970).

—— *Aid From OPEC Countries* (Paris, 1983).

—— *External Debt of Developing Countries* (Paris, 1982).

—— *Development Cooperation Annual Review* (Paris, various issues).

ROBERT S. PINDYCK 'The structure of world energy demand' (MIT Press: Cambridge, Mass., 1979).

—— 'Pouvoir de monopole sur les marches des resources non renouvables', in *Ressources Naturelles et Theorie Economiquw*, Gerard Gaudet and Pierre Lasserre (eds.) (Les Presses de L'Universite Laval, Quebec, 1986).

—— 'On monopoly power in extractive resource markets', *Journal of Environmental Economics and Management*, 14 (1987), 128–142.

KERRY V. SMITH, 'The empirical relevance of Hotelling's model for natural resources', *Resources and Energy*, 3 (1981), 105–17.

R. H. SNAPE 'Effects of mineral development on the economy', *Australian Journal of Agricultural Economics* (1977), 147–56.

ROBERT SOLOW 'The economics of resources and the resources of economics', *American Economic Review, Papers and Proceedings* 64, (1974), 1–14.

JOSEPH STIGLITZ 'Monopoly and the rate of extraction of exhaustible resources'. *American Economic Review*, 6.3 (1976), 655–61.

World Bank *World Development Report*'. Annual publication, Washington, DC.

—— *Crude Oil Imports: Commodity Trade and Price Trends* (Washington, DC, 1982).

Index